SNAKES

Biology, Diversity, and Behavior

David Gower,
Katie Garrett, and
Simon Maddock

Comstock Publishing Associates
an imprint of
Cornell University Press
Ithaca, New York

First published in 2000 by the Natural History Museum, Cromwell
Road, London SW7 5BD. This edition with updates published in 2023.

First published in the United States of America in 2023 by Cornell
University Press

© Natural History Museum, London, 2000, 2012, 2023

Librarians: A CIP catalog record for this book is available from the
Library of Congress.
ISBN 978-1-5017-7353-2 (paperback: alk. paper)

Designed by Mercer Design, London
Reproduction by Saxon Digital Services
Printing by Toppan Leefung Printing Limited, China

Front cover: South American tree boa, *Corallus hortulanus*. © Michael
D. Kern/Nature Picture Library.

Contents

Preface

Few animals evoke such strong emotions in humans as snakes – they are loathed, feared, admired, or even worshipped the world over. There can be hardly anyone without at least some perception of these distinctive and fascinating animals. But what exactly is it that inspires these feelings? For many it is undoubtedly because some snakes have deadly bites, but for most of us it is probably because they are difficult to relate to. How can an animal so long and thin move so gracefully without limbs, or swallow a meal several times larger than its head, and what is it that enables some snakes to survive without eating for months?

The first objective of this book is to answer some of these questions. Its other main purpose is to explore each of the main groups of snakes and convey some impression of the remarkable extent to which these animals have diversified during their evolution. Snakes inhabit almost every part of the globe where temperatures remain conducive to life for at least part of the year, including the open sea, and they have become specialized for living in a wide range of different environments. Some are adapted for life in water and never venture onto dry land, while others are found only in the treetops of forests, or spend much of their lives burrowing underground in dry, sandy deserts or moist tropical soils. Only in the coldest regions and on some islands (notably Ireland and New Zealand) are there no snakes at all. Equally varied and complex are the ways in which individual species live, and by focusing on aspects of their natural history we hope that this book will encourage the growing appreciation of snakes as an important group of animals that should be valued, admired and studied rather than feared and hated.

The history of this book deserves explanation because it includes a fine pedigree of books written by staff of the Natural History Museum, London. The origin lies with the 1963 book *Snakes* written by Hampton W Parker CBE (1897–1968), an expert herpetologist who was Keeper (head of the department)

OPPOSITE DeKay's brown snake, *Storeria dekayi*, a small natricine colubrid from the eastern parts of North America and Central America. It is often found in urban areas.

of Zoology between 1947 and 1957. This 1963 book was reincarnated by Parker in a modified version as *Natural History of Snakes* in 1965, and that version was expanded and revised by Alice G C Grandison in 1977 when it was published as *Snakes – A Natural History*. Grandison considered this 1977 version to be a second edition of Parker's 1965 book. Grandison (1927–2014) was head of the Natural History Museum's Herpetology Section, where she worked between 1951 and 1984. The 1977 version (and its antecedents) was clearly an influence on Peter J Stafford's 2000 book *Snakes*. Stafford (1961–2009) was employed by the Museum as a botanist, but moonlighted as a herpetologist and was widely known outside the Museum for his passion for and studies of amphibians and reptiles. The 2000 book was the first to use a very extensive set of photographs, but it also retained several of the figures that Brian C Groombridge had drawn for the 1977 version, and these figures are also included in this new version and its immediate predecessor. Two of the authors of this new edition (Gower and Garrett) revised and updated Stafford's 2000 book in 2012, under the same title, *Snakes*. All five previous versions have covered aspects of snake natural history as well as summarizing their taxonomic, ecological and morphological diversity, and we follow that model here. This new edition is a modified version of our 2012 book, updated to reflect advances in the knowledge of the natural history of snakes, and improvements in their taxonomy and classification. It owes much to its predecessors.

About the authors

DAVID GOWER is an organismal biologist whose research covers wide-ranging aspects of natural history such as evolution, taxonomy, biogeography, ecology, sensory biology, reproduction and conservation. His research is firmly rooted in museum collections and fieldwork, and his taxon expertise lies in caecilian amphibians and Triassic archosaurian reptiles as well as (mostly burrowing and aquatic) snakes. David has published more than 240 articles in scientific journals, including the description of more than 20 newly discovered species of snakes. He is a Merit Researcher in the Herpetology Section of the Natural History Museum, London and holds adjunct and associate positions at Central University of Kerala and University of Seychelles.

KATIE GARRETT is a British filmmaker and science communicator based in Virginia, USA. Her work focuses on the conservation of reptiles and amphibians, and her short documentaries have been published by online outlets such as National Geographic and bioGraphic Magazine. Katie's films have won awards at film festivals worldwide. With a degree in biology, she worked at the Natural History Museum, London, between 2009 and 2012 and has featured the work of NHM staff in several of her films, including her recent work with the Amphibian and Reptile Conservation Trust on the return of the northern pool frog to the UK.

SIMON MADDOCK is an organismal biologist specializing in reptiles and amphibians. His research covers a broad range of topics including ecology, evolution and conservation. To address these research and conservation activities he uses a multidisciplinary approach, incorporating elements of fieldwork, genetics, morphology, ecology, environmental sampling and capacity building, with most of this work focusing on British, Ecuadorian, Seychellois and Papuan species and habitats. Simon is a Lecturer in Ecology and Evolution at Newcastle University, UK, a Scientific Associate of the Natural History Museum, London and is also an Associate Member of the Island Biodiversity and Conservation Centre, University of Seychelles.

1 Structure and lifestyle

Approximately 4,000 living species of snakes have been discovered and formally described and named. They range in size from the burrowing thread snakes that may be as little as 10 cm (4 in) long, to giants such as the larger pythons and anacondas that grow to 7 m (23 ft) or more. Some specialized tree-living snakes are amazingly long and thin, whereas many of the vipers, boas and pythons are relatively short and heavy-set, and there is a whole range of different combinations in between.

Although they vary in size and shape, the features that collectively distinguish snakes as a group are clearly recognizable: the body is greatly elongated and highly flexible, there are no limbs or external ear openings, and the eyes have no eyelids. They differ further from other reptiles in lacking any sign of a shoulder girdle, forelimbs and a sternum (breastbone), and in the vast majority there are no vestiges either of a pelvis or hindlimbs. Only pythons, boas and a few other forms have remnants of pelves or hindlimbs, which lie entirely within the body or appear externally as small, horn-like claws at the base of the tail.

LEFT The ancestors of snakes may have looked superficially like the slow worm, *Anguis fragilis*, an Old World lizard that, like snakes, has lost its legs and acquired an elongate body. Slow worms differ from snakes in multiple details, for example in having eyelids.

OPPOSITE A grass snake, the natricine colubrid *Natrix natrix*, in Tuscany, Italy.

DISCOVERY OF 'NEW' SNAKE SPECIES

ABOVE The xenodermid *Stoliczkia vanhnuailianai* was discovered in forest in Mizoram, northeast India in 2021. A second specimen was found in the same year, but this remains a rarely encountered and poorly known snake.

LEFT The Indian shieldtail, *Rhinophis karinthandani*, was scientifically named only in 2020, though these previously misidentified specimens had been in the collections of London's Natural History Museum since 1879.

The number of species of snakes that is recognized by scientists as living in the world today fluctuates as snake classification is continually checked and rechecked. Sometimes, previously described species are removed from the global inventory because new evidence is found that they are not actually different to previously described species. Overall, however, the number of species of snakes recognized by scientists has been increasing substantially over the past few years. When the previous edition of this book was published, in 2012, there were approximately 3,370 species, but now there are more than 3,970, a net increase of nearly 18% in 10 years. The trend suggests that this number will continue to increase substantially for some time ahead.

Species of snake that are 'new' to science are typically discovered in one of three ways. First, they can be discovered during fieldwork. Second, they can be discovered in museum collections where specimens might have been stored for tens or hundreds of years without it being realized that they were examples of distinct species. These species are only 'new' in the sense that they have been discovered scientifically – sometimes these snakes were already known by non-scientists. Third, 'new' species can be elevated to species status from existing varieties or subspecies that were already previously known, or they can be resurrected from species names mistakenly removed from the inventory, if scientists find evidence that they do differ enough to warrant recognition as distinct species in their own right. Examples of two of these types of discoveries are shown here.

Taxonomy and classification: the scientific naming of snakes

To communicate effectively about organisms and their evolutionary relationships, humans have created a naming and organization system known as biological taxonomy and classification. In the widely used Linnean System (named after the eighteenth-century Swedish scientist Carl Linnaeus), each species is given a unique, two-part, combination name that lists the species designation (or epithet) after the name of the genus in which that species is classified, and these two words are written in italics with the genus name having an uppercase first letter. Thus, humans are *Homo sapiens*, and European adders are *Vipera berus*. In this book we typically refer to a particular species using its scientific name (in italics), and sometimes also mention an accompanying 'common' name. When writing about different species in the same genus within a single sentence or paragraph, the genus name is usually abbreviated to a single letter followed by a stop (period), for example the green and yellow anacondas, which are two species of the genus *Eunectes*, can be written about as *Eunectes murinus* and *E. notaeus*, respectively. Very rarely in this book, you will see three consecutive names in italics, these refer to a particular subspecies, for example the red-sided garter snake is *Thamnophis sirtalis parietalis*, which is a subspecies of *T. sirtalis*.

In this book we also often report the scientific family (and sometimes subfamily and superfamily) in which a particular genus and species is classified; these names are written in non-italic text with an uppercase first letter. For example, *Vipera berus* is a member of the subfamily Viperinae within the family Viperidae, i.e. *V. berus* is a viperine viperid (the adjectival versions of subfamily and family have a lowercase first letter). Names of animal subfamilies end with -inae, and families with -idae. Superfamily names end with -oidea, such as Elapoidea, to which elapids (members of the family Elapidae) belong.

As discussed further in Part Two, biologists prefer to give scientific names to unique evolutionary lineages of organisms rather than to groups of superficially similar organisms that might not all be each other's closest relatives. Routinely, evidence is discovered that corrects previous ideas about whether a population or other group of snakes is truly a distinct species, or scientists may propose that a particular species is better classified in another genus or subfamily or family etc., based on evidence of its evolutionary relationships. Thus, scientific classifications and the scientific names of particular snakes may change over time. For example, the sea snake *Hydrophis peronii* was originally described in 1853 as

Acalyptus peronii, which was modified in 1896 to *Acalyptophis peronii* once it was realized that the genus name *Acalyptus* had first been in use for a group of beetles! Then, in 2012, the species was transferred to *Hydrophis* (as *H. peronii*) in a new classification, based on DNA evidence that it was part of the *Hydrophis* lineage.

Although ongoing change can make scientific names confusing, they are more precise than 'common' names when it comes to communicating about the many different species of snakes across the world. Common names typically vary across different languages, or there may be multiple common names in use for a single species, or multiple species may be referred to by a single common name. Additionally, many common names are used for groups of snake species that do not represent unique lineages of snakes. For example, at least two not especially closely related groups of nocturnal snakes with large eyes have been called 'cat' or 'cat-eyed' snakes: the Asian colubrine colubrids (genus *Boiga*) and close relatives, and several neotropical dipsadine colubrids. 'Garter' snakes are very different animals in Africa (the highly venomous elapid *Elapsoidea*) and North America (the natricine colubrid *Thamnophis*). And, without further clarification, 'racers' might refer to some of the fast-moving, generally ground-dwelling snakes in the Colubrinae, Dipsadinae or Pseudoxyrhophiidae.

The origins and fossil record of snakes

Snakes (classified as the suborder Serpentes) are members of the order Squamata (i.e. they are squamates). Squamates comprise all living reptiles except for crocodiles and alligators (order Crocodylia), tortoises and turtles (order Testudines) and the tuatara (order Rhynchocephalia). Thus, squamates include all living snakes and lizards, the latter including some superficially snake-like groups such as limbless skinks, slow worms, and worm lizards (amphisbaenians). Snakes' closest living relatives are the anguiform lizards (monitor lizards, beaded lizards, knob-scaled lizards, galliwasps, slow worms, glass lizards and alligator lizards) and/or iguanians (iguanas, chameleons, anoles and agamas). Whatever the precise relationships among snakes and other living squamates, it is clear that snakes evolved from lizard ancestors.

The many divergent features of snakes, including their elongate and limbless body and unusual eyes, have resulted in two main competing explanations for snake origins. The dominant hypothesis is that snakes underwent a burrowing phase in their immediate ancestry or early history. The supporting evidence includes the observation that several lineages of lizards (e.g. slow worms, dibamids, amphisbaenians, pygopod geckos) have become elongate and have lost or substantially reduced their limbs, and these generally burrow in compost, sand or soil. Some of the peculiarities of snake eyes are the result of evolutionary loss of some ancestral features, and this is also known to have occurred to varying degrees in other burrowing vertebrates. Additionally, some of the oldest fossil snakes have features that suggest they might have been burrowers. The main competing hypothesis is that snakes passed through a profoundly aquatic (and marine) phase in their immediate ancestry. Some scientists have claimed that snake eyes are similar in several features to those of other aquatic vertebrates. The main supporting evidence for the marine hypothesis, however, comes from similarities between snakes and extinct marine lizards. There are also several snake fossils that retain hindlimbs and that were fossilized in marine sediments deposited in the Cretaceous period (145 to 66 million years ago). Supporters of

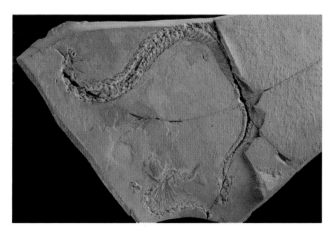

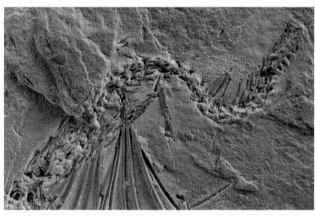

LEFT AND BELOW Fossil of the extinct marine snake *Eupodophis descouensi* preserved in Cretaceous rocks from Lebanon, nearly 100 million years old. The left image shows the whole specimen preserved in a slab of rock. The image below shows a close-up of part of the vertebral column and a small but well-formed right hindlimb, with part of the ankle and the foot missing.

this marine hypothesis argue that elongation of the body and reduction of limbs arose as an adaptation to swimming in water rather than burrowing on land. The marine hypothesis has been challenged by evidence that the Cretaceous limbed snakes are not close to the immediate ancestry of living snakes. Instead, they are probably embedded more deeply within the evolutionary tree of living snakes, so that their limbs might have partly re-evolved from small stubs such as are retained in, for example, living pythons.

Although the marine, limbed Cretaceous snakes are known from some spectacular and complete fossil specimens, much of the fossil record of snakes is patchy and fragmentary. Many snake fossils consist of no more than isolated vertebrae. Interpreting these fossils can be difficult because even vertebral variation within and among individuals and species of living snakes is not well known in detail. It is particularly tricky to identify the earliest member of any major group of organisms based on fragmentary fossils, and this is true also for snakes. Geologically the oldest snake fossils reported so far are from Cretaceous rocks in North America and North Africa, dating to about 100 million years ago. It is likely that snakes diverged from their closest-living, lizard relatives much earlier than this, perhaps in the Jurassic, but convincing fossils have yet to be found. Some much older fossil finds from Europe (to more than 160 million years old) have been claimed to be more closely related to snakes than to any living lizard group, but most palaeontologists do not consider these fossils to be snake-like enough to be sure.

RIGHT A fossil vertebra of the extinct giant snake *Titanoboa cerrejonensis*, excavated from rocks approximately 60 million years old in Colombia. This species reached lengths of at least 13 m (42½ ft), and the fossil specimen shown here dwarfs a similar vertebra of a 5.2 m (17 ft) long modern day green anaconda, *Eunectes murinus* (also see p.89).

SNAKES IN HISTORY AND MYTHOLOGY

ABOVE A black snake is incorporated into the fabric of the holiest Yazidi temple of Lalish, Iraqi Kurdistan.

Snakes have long been the subject of stories and have been worshipped and revered throughout human history. In some cultures, it is a fundamental belief that snakes are the givers of life. For example, some Indigenous Australians believe that in the Dreaming (a period relating to the origins of the world) the 'Rainbow Snake' is responsible for creating water courses and, by connecting waterhole to waterhole in the form of a rainbow, preventing permanent waterholes from drying up. Other cultures also feature a 'Rainbow Snake' in their oral traditions; in some regions of West Africa, it is believed that the 'Rainbow Snake' held up the sky whereas in others that it causes flooding and earthquakes.

In Ancient Egypt snakes were portrayed as good through the deity Wadjet and by being associated with the Sun God Ra. The gold cobra Uraeus was the symbol of Egyptian sovereignty, being displayed on the foreheads of the Egyptian Royal Family and goddesses. Additionally, in Ancient Egypt, snakes were also portrayed as evil, with Apep being the deity of chaos. In Ancient Greek mythology the Hydra was a nine-headed sea serpent that Hercules had to slay as one of his twelve labours, and Medusa was a Gorgon with snakes replacing her hair — people looking into her eyes would turn to stone. Religions that are still around today have snakes mostly depicted as symbols of evil, the most notable being the story of a serpent persuading Adam and Eve to eat fruit from the forbidden tree in the Garden of Eden, which appears in Christian, Jewish and Islamic religious texts.

Snakes have been used as symbols for millennia. The circular image of a snake biting its own tail (Ouroboros) represented the continuity of life and death in Ancient Egypt and Greece, with similar depictions in other cultures. In the late 1800s, the chemist August Kekule claimed that a vision of the Ouroboros inspired him to work on equations to prove the circular structure of the benzene molecule.

Snakes are still depicted as the symbol of modern medicine. The Caduceus symbol is a staff entwined by one or two snakes, which is the emblem for the Greek god of medicine Asclepius and Roman Aesculapius. The Aesculapian snake, the colubrid *Zamenis longissimus*, derives its common name from the belief that they are the snakes depicted on the Caduceus.

LEFT The Staff (or Rod) of Asclepius. In Greek mythology the Staff was wielded by the god Aesculapius, who was associated with healing. The depiction of the Staff is commonly used today as a symbol for health care.

Other fossils have had an important influence on the understanding of the evolutionary history of snakes. Fossilized viper and elapid fangs are known from German rocks of approximately 20 million years old. The oldest-known scolecophidian remains are from rocks approximately 65 to 70 million years old in North and South America. Evidence for viviparity has been discovered in the form of fossilized embryos within an approximately 50 million years old female boa specimen from Germany. One fossil specimen of the extinct, mostly Cretaceous snake family Madtsoiidae, was found preserved in a sauropod dinosaur nest. Other madtsoiids reached lengths of 10 m (34 ft), longer than any living snake, but these were still greatly outsized by the perhaps 14 m (46 ft) long Cretaceous *Titanoboa*. Future fossil discoveries are likely to have a dramatic impact on our understanding of the timing and pattern of the evolutionary diversification of snakes, especially when studied using modern imaging techniques, such as micro computed tomography (microCT scanning), which enable palaeontologists to use X-rays to visualize exquisite details in fossils that are crushed and/or still embedded in rock.

Anatomy

Over the course of evolution, elongation of snakes' bodies has necessitated the modification and rearrangement of their internal organs. Most of the main organs themselves are present and not very different from those of humans and other vertebrates, but they have changed so much in shape and position that it can be difficult to recognize them at first glance.

The lungs in particular have undergone considerable modification. Some snakes have two lungs, but in these cases the left is always smaller than the right. In some pythons the left lung can be up to 85% of the length of the right lung, but in the majority of snakes the left lung has been either lost completely or become greatly reduced, as little as 1% of the length of the right in some snakes in the family Lamprophiidae. However, in some aquatic snakes, such as the Asian and Australasian file snakes (*Acrochordus* species), the right lung is particularly large, extending backwards for nearly the entire length of the body, likely to enable them to hold their breath for long periods spent underwater. A possibly additional respiratory area in many snakes is provided by a special modification of the windpipe, known as the tracheal lung.

In most vertebrates, paired organs such as the kidneys and gonads usually lie in the same position on either side of the body, but these have become staggered in

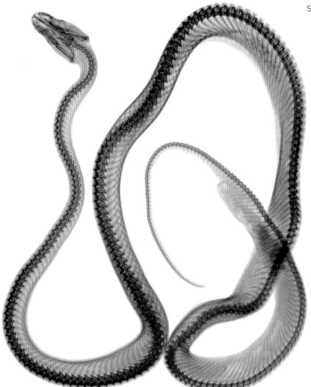

LEFT A snake's skeleton is comprised mostly of vertebrae and pairs of ribs. There is no sternum, shoulder girdle or forelimbs, although in some, such as this python, vestiges of the pelvis and hindlimbs remain.

BELOW The paired hemipenes of a male North American rat snake, everted from the vent at the base of the tail. Biologists often use the details of size, shape and ornamentation of these structures to help distinguish different species.

THE ANATOMY OF A MALE SNAKE

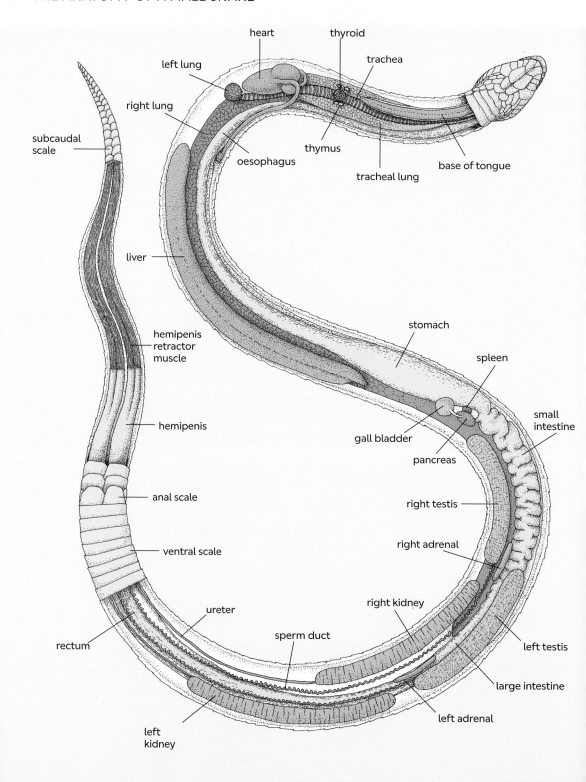

the elongate body of snakes. Among other organs affected by the radical change in body shape is the stomach, which has become greatly enlarged and, in some snakes, accounts for more than one-third of the total length of the body. There is also no urinary bladder; nitrogenous waste is voided not in a solution of urea, as in humans and almost all other mammals, but in a semi-solid state as uric acid, as occurs in lizards and birds. The heart has three chambers, as opposed to four in humans, while the male copulatory organ is a pair of structures, called hemipenes (as also found in lizards), variously and often spectacularly adorned with frills, spines and other ornamentation. Female snakes similarly resemble female lizards in having paired hemiclitores, rather than the single clitoris of turtles, crocodilians and mammals.

SKULL AND TEETH

The most distinctive feature of the skull in the majority of snakes is its remarkably flexible construction. With some exceptions (e.g. pipe snakes), most of a snake's skull bones are movably connected to each other and attached only loosely to the braincase, so that much of the head is capable of being stretched and distorted in many directions. The two halves of the lower jaw are not fused at the front into a solid 'chin' but are often separated by elastic ligaments that allow them to be forced apart when the snake is swallowing large prey. The suppleness of the lower jaw is further enhanced by a joint in the middle of each of its two halves that enables the jaws to be flexed outwards. The lower jaws connect to the skull via the quadrate bones. In many snakes the quadrates are elongate and their lower ends can swivel outwards, allowing the rear ends of the lower jaws to be forced apart and large prey to pass into the throat. Despite popular beliefs, snakes thus do not dislocate their jaw joints when feeding.

The teeth of snakes are typically thorn-shaped and curved backwards to varying degrees. However, there is much variation on this theme in different snake lineages, the details of which generally remain little studied and poorly understood. As well as the teeth lining each of the upper and lower jaws, there are normally two further rows on the roof of the mouth (the palatine and pterygoid bones). Snake teeth are replaced alternately and repeatedly throughout life. Stiletto snakes, some colubrids, elapids (mambas, cobras etc.) and vipers have one or more pairs of enlarged teeth that are specially modified for injecting venom, and the teeth of other species are likely adapted for catching and ingesting particular kinds of prey. Those of the Central American neck-banded snake, the colubrid *Scaphiodontophis annulatus* for example, are slightly flattened at their tips for grasping the smooth, hard-scaled bodies of skinks on which it largely feeds.

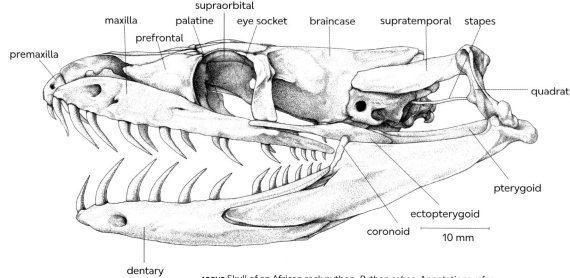

supraorbital
maxilla   palatine   eye socket   braincase   supratemporal   stapes
prefrontal
premaxilla
quadrat
pterygoid
ectopterygoid
coronoid
10 mm
dentary

ABOVE Skull of an African rock python, *Python sebae*. Annotations refer to bones mentioned in this book. Note in particular the coronoid bone, a primitive feature of the lower jaw that has been lost in many snake lineages.

The teeth of this snake are also hinged on flexible ligaments that enable them to be locked into a backward-pointing position when prey is being swallowed, thus preventing it from struggling free.

The presence and position of different types of fangs used to be of prime importance in previous classifications of the major lineages of snakes. It is now known that evolutionary changes among the main different types of dentition have occurred several times. Thus, even closely related snakes might not necessarily share the same types of dentition, so teeth are not always a reliable guide to evolutionary relationships. Recently, detailed modern studies of developmental biology have demonstrated that in all snakes with fangs these enlarged teeth develop first at the rear of the upper jaw, and that in front-fanged snakes (e.g. vipers, elapids) they migrate forwards during development. Thus, front fangs of different snake lineages probably have a common developmental origin, but the anterior migration has evolved independently in different groups.

Developmental studies have also shown that tubular venom-delivering fangs of elapids and vipers grow their tubes not by a progressive infolding of the long edges of the teeth, but by adding more tooth material around the base of an incipient tube that is present when the teeth first form. Snakes with tubular fangs have muscles associated with their venom glands. These muscles squeeze

TYPES OF DENTITION IN SNAKES

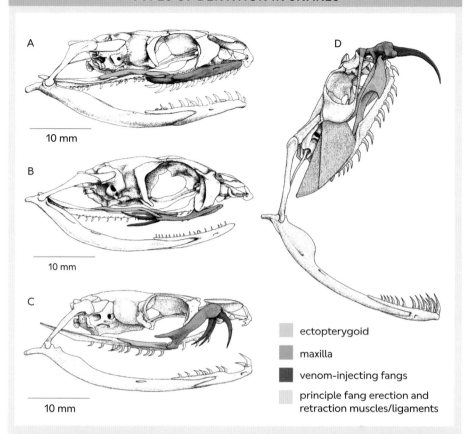

A

10 mm

B

10 mm

C

10 mm

D

ectopterygoid

maxilla

venom-injecting fangs

principle fang erection and
retraction muscles/ligaments

Features shaded represent the main components of the biting mechanism; note in particular the relative sizes of the maxillary bone.

A No enlarged fangs (aglyphous). Includes thread and blind snakes, pipe snakes, sunbeam snakes, boas, pythons and many non-venomous caenophidians.

B Enlarged, rear-mounted fangs usually preceded on the maxilla by several smaller, unmodified teeth (opisthoglyphous). Includes many non-elapid and non-viperid, venomous caenophidians, including many colubrids.

C Enlarged, forward-mounted fangs that are non-erectile (proteroglyphous). Includes cobras, mambas, coral snakes and all other elapids.

D Greatly reduced maxilla bone with enlarged, forward-mounted fangs that are erectile and capable of being pivoted independently (solenoglyphous); when not in use the fangs fold back along the upper jaw. Includes stiletto snakes (atractaspidine atractaspidids) and all vipers, though in these two groups the nature of the articulation between the maxilla and the prefrontal is different. All fangs are on the upper jaws, and these enlarged teeth are also tubular or grooved to facilitate the delivery of venom into prey.

the venom out at high pressure, necessary to move it forcibly through the narrow bore of the needle-like teeth, and into the flesh of prey. It has been argued that tubular fangs evolved only once within snakes (and were subsequently simplified or lost in several lineages) rather than evolving independently at least three times – in elapids, vipers and stiletto snakes.

SKIN, SCALES AND MOULTING

As in other reptiles, the entire surface of the body of almost all snakes is covered with scales (for an exception see p.96). On the head these may be large (sometimes termed 'shields') and arranged in a symmetrical pattern, or small and irregular. Those on the upper surface and sides of the body (dorsal scales) are usually small and regular, sometimes with a horizontal ridge across the centre (keeled). In contrast, the scales on the belly (ventral scales) are typically large and broad and extend crosswise in a single line between the head and base of the tail. In some snakes the scales beneath the tail (subcaudals) are also arranged in a single row, while in others they are paired. In many snakes the scales of the snout and chin are covered with microscopic tubercles that appear to be tactile mechanoreceptors, and similar tubercles occur near the base of the tail in some species, including sea snakes and New World coral snakes. These sensory tubercles often differ in size, distribution, and abundance between males and females of the same species, suggesting they are important in reproduction.

The outermost layer of a snake's skin consists of a continuous sheet of keratin, a rather inflexible material found also in human fingernails and hair. Derived from dead cells, this layer must be shed every so often to allow for growth and repair. The frequency with which this shedding, or sloughing, occurs depends primarily on the rate of growth, and young healthy snakes, which grow more quickly, slough their skins more often.

In the initial stages of this process, the skin loses its usual fresh appearance and turns milky white, due to the secretion of a lubricant beneath the redundant upper layer: this is especially pronounced over the eyes. It persists for several days and may take up to three weeks before it finally clears, during which time the snake is often unable to see properly and will usually remain hidden away. Many snakes also refuse to feed during this period, and if disturbed may become defensive. A few days after fully regaining its sight, the snake becomes restless and begins to rub its head against the ground, stones, or other rough surfaces until the old skin separates at the tip of the snout and along the jawline, from where it then peels back over the rest of the body as the snake moves around.

LEFT Individual scales of a snake's body are separated by interconnecting skin that is generally concealed beneath. This allows for the considerable flexibility needed for movement or consuming large meals. The broad scales on this ground boa from Madagascar, *Acrantophis dumerili*, are the 'ventrals' along the midline of the belly.

In the process, the old skin is turned inside out. Sea snakes that live in open water cannot always find suitable rough surfaces, but some of these species have ingeniously solved the problem by coiling themselves into knots and rubbing one part of the body against another.

The very fine texture (micro-ornamentation) of the surface of scales is adapted to serve different purposes depending on the ecology of the snake in question. For example, ventral scales of ground-dwelling snakes are often shiny and seem to be smooth, probably partly to reduce friction as they slide their bellies along the substrate, but also because the underside of the snake is rarely seen and so even if it is shiny it will not alert predators or prey to the snake's presence. For the same reason, many snakes have much duller and light-absorbing or light-scattering (rather than reflecting) dorsal scales. Many soil-dwelling burrowing (fossorial) snakes have exceptionally iridescent scales. This iridescence probably serves no immediate function, but it appears to be an accidental consequence of evolving scales that are covered in very regularly spaced, microscopic ridges – features that serve to repel moisture and dirt.

Although scale microornamentation can lead to structural colour, such as iridescence, most snake colour patterns are caused by colour pigments in the skin. Many snake species have colour patterns that make them cryptic in their environment, yet some species are brightly coloured to warn predators that they are venomous. Several non-venomous species have also adapted bright colouration to mimic their venomous counterparts (see p.38).

AN EMERGING THREAT: SNAKE FUNGAL DISEASE

ABOVE A massasauga rattlesnake, *Sistrurus catenatus,* with an *Ophidiomycosis* infection affecting the front of its head.

Snakes are vulnerable to many different diseases, but one in particular has recently come to scientists' attention as a potential conservation concern. Snake Fungal Disease or *Ophidiomycosis* was first recorded in rattlesnakes in the USA in 2006 where it was found to infect massasaugas (the pit viper *Sistrurus catenatus*), and the timber rattlesnake, *Crotalus horridis*. Since then, it has been reported in snakes throughout the eastern USA, Canada, parts of Europe and Australia, and is currently present in 11 countries and found to infect 62 different snake species.

Snake Fungal Disease is caused by infections of the fungus *Ophidiomyces ophidiicola*, which results in skin lesions and dermatitis and can be fatal. It has also been found to affect behaviour, thermoregulation and reproductive physiology, which further threatens individuals' health.

The disease has been widely studied because it is thought to be causing localized population declines. There is, however, currently debate as to whether it is a new, emerging disease, or whether it is an endemic disease that we have recently discovered. Either way it is adding pressure to snake populations that are already threatened by multiple stressors (see p.123). Studies have shown that captive snakes with the fungal disease have a higher mortality rate than wild individuals; one study showed over half of infected captive snakes died, as opposed to approximately 10% of wild counterparts.

Wild infection rates seem to vary considerably depending on species and geographic region. With so many pressures on snake populations worldwide it is an added threat that is being taken very seriously by conservation biologists and many studies are currently underway to understand more about this disease.

Senses

Snakes have an acute sense of chemoreception (taste and smell). Odorants (chemical compounds having a smell) are detected using two independent systems, the 'nose' (olfactory system) and the vomeronasal (or Jacobson's) organ located in the roof of the mouth. The nose smells odorants drawn in through the nostrils, while the vomeronasal organ senses odorants that are passed to it physically by the tongue as it flicks in and out of the mouth. There are a pair of vomeronasal chambers in the roof of the mouth, with odorants passed to them by the two tips of the forked tongue. Although the two systems are somewhat disconnected and use different signalling pathways, it seems that sensing by olfaction is important for increasing the rate of tongue flicking. So acute are snakes' sense of chemoreception that they can detect even the faintest scent trails left by prey – or another snake. The specialist ant- and termite- hunting scolecophidian snakes, for example, can accurately follow an ant trail even a week after the trail has been made.

LEFT As well as using their nose, snakes detect odour by using their tongue to transfer odorants to the vomeronasal (Jacobson's) organ in the roof of the mouth. The tongue may also have a tactile function. The western tree snake, *Imantodes inornatus,* an arboreal dipsadine colubrid, photoghraphed in Costa Rica.

SIGHT

Snakes' eyes are different from those of other animals. They lack eyelids, and most species instead have a fixed, transparent 'spectacle' protecting the eye. In humans and most other vertebrates, focusing is achieved with special muscles that change the shape of the lens, whereas in snakes there are no such muscles and images are focused instead mostly by moving the hard lens physically towards or away from the retina. Snake retinas are also interesting in that, as a group, they have a greater diversity of rod and cone photoreceptor cell types than other vertebrates, though the functional and evolutionary significance of this is not well understood. Biologists still know very little about how snakes see, but it seems that for many species their eyes will be used to detect movement rather than forming very detailed images: many snakes hunting by sight might fail to recognize a potential meal unless it begins to move, even if it is only a few centimetres away. However, some snakes are highly visual predators with excellent eyesight, such as the Montpellier snake (the psammophiid *Malpolon monspessulanus*), and species of the colubrine colubrid genera *Ahaetulla* and *Chrysopelea*. These all have pigmented lenses that prevent UV light from reaching the retina, which allows for sharper image formation, and they will move their heads to visually track distant objects, even aeroplanes passing overhead.

ABOVE Snakes do not have eyelids. Instead the eyes are covered by a transparent protective cap (the brille), as in this neotropical racer, the colubrine colubrid *Drymoluber dichrous* photographed in French Guiana.

HEARING

Snakes have neither an external ear nor an eardrum, although the sense of hearing in many species appears to be otherwise developed and functional. Snakes are best suited to detect ground-borne rather than air-borne sound vibrations, which are picked up by the bones of the lower jaw and transmitted to the inner ear via a delicate, bony rod known as the stapes. However, research has suggested that the ear of many snakes may be more sensitive to airborne or aquatic sound than has previously been assumed, and some herpetologists (people who study reptiles and amphibians) think the lung may also act as a resonator.

OTHER SENSES

Rattlesnakes and other pit vipers, together with pythons and boas, are exceptional among vertebrates in having sense organs that detect infrared radiation. These organs in pit vipers are in a special pit on the side of the face, while in pythons and boas they are within, between or behind scales along the margins of the mouth. The thermal information is combined in the brain simultaneously with visual images from the eyes, allowing these snakes to form rich and detailed 'pictures' of their environment. Vipers are able to detect radiation emitted by 'warm-blooded' prey at distances of up to 1 m (3¼ ft). The pit organs of many pit vipers might even be able to detect very small differences in temperature between 'cold-blooded' animals such as frogs and the environment, and thus locate them in the dark. Just as remarkable, some species of the sea snake genus *Aipysurus* have light-sensitive tissues in their tails. These species use this ability to ensure that their tails are not left exposed when it is hiding among crevices during the day (see p.147).

Temperature regulation

Snakes are considered ectothermic animals, meaning that they rely almost entirely on external, environmental sources of heat to regulate their body temperatures, unlike endothermic mammals which can regulate their internal temperatures by metabolic processes. To achieve their optimum 'working' temperature snakes expose themselves to whatever source of heat may be available, such as direct sunlight or a sun-warmed rock. This is why snakes are often seen basking in the sun. Should they begin to overheat, they cool down by moving into shade, burrowing underground, or immersing themselves in water. They must keep their body temperature within a range of about 4–38°C (39–100°F) depending on the species; if their temperature falls or rises only a few degrees either side of these levels, they can die.

HOW SNAKES MOVE

ABOVE A horned adder, the viperid *Bitis caudalis*, 'sidewinding' over a desert sand dune in Namibia.

How snakes move is one of their most intriguing features. Part of the answer lies in their large number of vertebrae and ribs, which provide the flexibility needed for limbless locomotion. Whereas humans have 43 vertebrae, some species of snakes have over 500. The vertebrae of snakes are among the most elaborate and complex found in any backboned animals, with a wide range of structures for supporting the ligaments and tendons that enable them to move in their characteristic 'slithery' manner.

Most snakes proceed with continuous side to side undulations formed by the natural progression of the body as it follows in line behind the head, where forward movement is achieved by the animal using each of the resulting S-shaped loops to push against irregularities in the ground. If the ground over which the snake is moving is unstable, as with loose, shifting sand, or there are no stones, twigs or other adequate 'footholds' which its body can push against, its ability to move is seriously impaired.

Many of the larger, stouter-bodied snakes, such as pythons, boas and ground-dwelling vipers, can also employ 'rectilinear' locomotion, whereby they move along with the body extended in almost a straight line. It used to be thought that this was achieved by effectively 'walking' on the tips of their ribs, but the ribs stay in position and, instead, groups of ventral scales are raised, moved forward and then placed back down on the ground and the body is drawn forwards over them. Another form of locomotion used by heavy bodied snakes and also by many burrowing and climbing forms is the 'concertina crawl', in which the animal advances by reaching ahead and pushing its body against the ground, a branch or some other firm point of anchorage, from where it is then able to pull the rest of its body up behind.

In sandy deserts or other areas where there is no firm ground, several kinds of snakes have developed a modified form of progression known as sidewinding. They throw the body into a succession of S-shaped coils and, instead of the more usual 'slither', take a succession of obliquely oriented 'steps' over the ground, during which the body is in contact with the surface at only two points at any given moment. The best-known exponents of this type of movement are desert-living vipers.

At times when prevailing conditions make it impossible for snakes to reach or regulate their body temperatures, they escape by retreating to a state of dormancy (aestivation). During the freezing conditions of winter in North America, for example, garter snakes, hibernate in underground dens, sometimes with thousands of individuals, while in some tropical regions that have long, hot, dry seasons, many snakes will secrete themselves in a cool burrow or beneath the bark of a tree and aestivate to avoid the effects of dehydration. Despite their reliance on external heat sources and the often cool feel to their bodies when handled, a snake's body temperature can be as high as our own after only a short time basking in the sun, and it is thus misleading to refer to these animals as 'cold-blooded'.

Some snake species have become so well adapted to absorbing heat from the sun that they can tolerate sub-zero ground temperatures as long as the sun is out. For example, the most northerly species of snake in the world, the adder, *Vipera berus*, can be seen basking on snow in Scandinavia. These adders achieve this feat by being entirely black (unlike their British counterparts) which helps absorb heat better than paler colours.

Feeding and diet

All snakes are carnivorous. Under natural conditions they feed more or less exclusively on living prey. A few will occasionally eat carrion (e.g. the diet of some North American cottonmouth vipers, *Agkistrodon piscivorus*, includes fish regurgitated by parent seabirds attending nestlings), but this tends to be the exception rather than the rule. Many of the more slender and agile snakes forage actively for prey, while others, such as boas, pythons and vipers, are mostly ambush hunters that lunge from a hiding place at passing animals. The predatory strike of some snakes is often too fast to follow with the human eye, and in some species is delivered with such force that much of the body may be thrown forwards. A number of viper species and several other ambush hunters have contrastingly coloured tails to lure prey within striking range (see p.102), and the African vine or twig snake (the colubrid *Thelotornis capensis*), is said to use its brilliantly coloured red tongue to attract prey in much the same way. Perhaps the most complex tail lure of any snake is that of the spider-tailed horned viper, *Pseudocerastes urarachnoides*, which looks very spider-like and is used to lure insectivorous birds within striking range (see pp.109-110).

ABOVE Spotted python, the Australo-Papuan *Antaresia maculosa*, constricting a rat. Coils thrown around the body of the prey prevent it from inhaling, and death is brought about by asphyxiation.

SWALLOWING PREY

Snakes normally swallow their prey whole, although there are a few instances where they may first discard some body parts. Thread or blind snakes, for example, may break off the heads of termites before swallowing them, and white-bellied mangrove snakes (the homalopsid *Fordonia leucobalia*), twist the legs off a crab if the animal itself is too large to swallow in one piece (see p.121).

The size of prey eaten by some snakes is truly amazing. Green anacondas, *Eunectes murinus*, and several species of large python are well known for their ability to consume deer, pigs and occasionally even humans, while small African egg-eating snake (the colubrid genus *Dasypeltis*), with heads scarcely wider than a fingernail, can swallow a hen's egg (see p.172). Some vipers have been known to ingest meals exceeding 150% of their own body weight. Such incredible feats of swallowing are made possible by the development in many snakes of an enormously distensible and flexible mouth (see p.19). Because they have no sternum (breastbone), the ends of the ribs can also separate widely to allow large prey to pass into the stomach.

Many non-venomous snakes begin swallowing as soon as they have secured a firm grip on their prey, whereas burrowing asps, elapids, vipers and some rear-fanged snakes first produce paralysis or death in their victim with venom. Others, such as the pythons and boas and other non-venomous snakes, first immobilize prey by suffocating it by constriction within the coils of their bodies.

A large-gaped ('macrostomous') snake that feeds on large diameter prey ('macrophagous') usually starts swallowing as soon as it has adjusted its prey into a head-first position. With the jaws stretched over and around the prey, the two rows of palatal teeth in the roof of the mouth repeatedly ratchet the prey backwards into the throat, eased along during the process by a lubricating coat of saliva. Wave-like contractions of the oesophagus then push the prey animal down into the snake's stomach. The snake may take up to four hours or so to swallow large, bulky prey, with most of this time taken to stretch the jaws around the prey. During swallowing a snake can keep breathing even while its mouth and throat are full by extending its airway to the outside of its mouth.

ABOVE Spotted cat-eyed snake, the dipsadine colubrid *Leptodeira septentrionalis*, swallowing a frog. The ability of many snakes to swallow large prey is made possible by the development of an enormously distensible mouth and elastic skin.

BEHAVIOURAL SECRETS OF SNAKES

ABOVE Australian water python, *Liasis fuscus*.

LEFT Fitting a radio telemetry transmitter into an anaesthetised Russell's viper, *Daboia russelii*.

Owing to their elusive habits, the inherent difficulties of observing them in the wild, and insufficient research effort, the natural lives of many snakes remain largely unknown. One of the few species to have been studied in any great detail is the water python, *Liasis fuscus*, an inhabitant of the seasonally flooded grasslands and billabongs (backwater pools) of northern Australia. By implanting some water pythons with miniature radio transmitters and monitoring their movements (a method known as radiotelemetry) over several years, biologists have been able to build up a detailed picture of their natural history.

Water pythons occupy different habitats and feed on different prey depending on the time of year. During the dry season, they spend the days hidden in dense reedbeds, emerging at dusk to feed more or less exclusively on rats. This pattern of activity continues throughout the dry season until the rains arrive and the rats abandon their homes to escape the rising water levels. At this time the snakes move out into the spreading floodwaters, where they become more or less completely aquatic and switch to feeding on water birds and their eggs.

Researchers have also recently started using radiotelemetry to study the poorly known natural history of potentially dangerous venomous snakes in India. The king cobra, *Ophiophagus hannah*, and Russell's viper, *Daboia russelii*, share habitats with people and domesticated animals across much of their range, so knowing where, when and how they spend their time should help efforts to both conserve these species and reduce the number of snakebite casualties. New discoveries made by this research include that king cobras have a home range of approximately 6–8 sq km (2¼–3 sq miles), and that they will sometimes eat dead as well as live snakes, including members of their own species.

DIFFERENT DIETS

Whereas many snakes eat a wide range of prey, others have highly specialized diets. Some feed almost exclusively on lizards, birds or rodents, and there are many that eat nothing but frogs. Among the most unusual dietary specialists are scolecophidian snakes (see p.57), which feed almost exclusively on ants, termites and their pupae, and the African egg-eating snakes (see p.172), snail-eaters (see p.100, p.177), Central American scorpion-eaters, and some sea snakes that eat only fish eggs (see p.149). Other kinds of prey exploited by specialist feeders include reptile eggs, earthworms, centipedes, bats, crabs, salamanders and amphisbaenians (burrowing, snake-like lizards), and a surprisingly large number of species have a diet that consists chiefly of other snakes.

Some species, such as the coastal taipan (the elapid *Oxyuranus scutellatus*), of Australia and the black mamba (the elapid *Dendroaspis polylepis*) of Africa, live on one or two kinds of prey throughout their entire lives, whereas the food preferences of many others vary according to their age and size, the time of year, or geographic location. Juvenile striped swamp snakes (the natricine *Regina alleni*), of North America for example, eat shrimp and dragonfly nymphs, whereas adults feed on crayfish. Water pythons, *Liasis fuscus*, in Australia subsist on floodplain rats during the dry season and change to a diet of water birds and their eggs in the rainy season (see p.32), while Mexican parrot snakes (the colubrine *Leptophis mexicanus*), eat frogs on mainland Central America and mostly lizards on some of the offshore islands.

A well-known feature of snakes is the ability of some species to survive without food for long periods. Pregnant female anacondas in the wild, for example, do not normally eat for the entire six to eight months of gestation, and black tiger snakes (the elapid *Notechis ater*), on Mount Chappell Island in the Bass Straight off southern Australia may feed for only a few weeks per year when their principal prey, the chicks of muttonbirds, are available. The longest recorded interval during which a snake has survived without food is held jointly by a green anaconda, *Eunectes murinus*, and an African rock python, *Python sebae*, both of which apparently refused food in captivity for three years before eating again. Under normal circumstances, snakes maintain themselves in a continuous state of readiness for feeding, but if food becomes scarce and there are lengthy intervals between meals, the internal organs of some species reduce to a state of temporary suspension (see p.83). Many snakes that feed on large prey may eat only once per week to once per month.

Venom

According to their principal clinical effects, most venomous snakes have venoms that are primarily neurotoxic (attacking nerve tissues and interfering with the transmission of nerve impulses), haemotoxic (directed towards the blood and circulatory system) or cytotoxic (causing cell damage and death). Neurotoxic venoms are characteristic mainly of elapid snakes, such as cobras, mambas and coral snakes, bites from which affect the central nervous system and typically lead to death by muscle paralysis and respiratory failure. The haemotoxic venoms of vipers and many Australasian elapids can disrupt blood clotting which in humans often leads to a haemorrhage (excessive bleeding) but in small prey animals causes a stroke. Cytotoxic venoms cause tissue destruction and they often work in parallel with haemotoxic action, leading to internal and external tissue destruction and potential death. Each snake venom consists of thousands of proteins, often having properties from all three major venom categories. Bites by neotropical rattlesnakes (the viperid *Crotalus durissus*), for example, cause breathing problems as well as symptoms more typical of viper envenomation, while victims of bites from the black-necked spitting cobra (the elapid *Naja nigricollis*), often suffer serious local tissue damage.

ABOVE Russell's viper, the viperid *Daboia russelii*, being 'milked' for the production of snakebite antivenom.

ABOVE Vipers, such as this bush viper, *Atheris ceratophora*, from Tanzania, typically have venoms that are haemotoxic in nature.

The toxicity of venom is normally measured by calculating the LD50 (50% lethal dose), which is the amount required to kill half of the laboratory animals (normally mice) into which it is injected. These numbers provide a general indication of how dangerous the venoms of different snakes may be, but in specific terms they show little correlation with the actual clinical danger of any given species to humans after a bite. Calculations of the LD50 value in viperid and elapid snakes such as saw-scaled vipers (*Echis* species), bushmasters (*Lachesis* species), kraits (*Bungarus* species) and even the black mamba, *Dendroaspis polylepis*, for example, look relatively innocuous on paper, whereas researchers know from clinical experience that these are among the most dangerous snakes, with high fatality rates. Furthermore, snake venoms differ considerably in their effect on different animals. The African meerkat, for example, which weighs only about 600 g (21 oz), is, weight-for-weight, one thousand times more resistant to the venom of the Cape cobra, *Naja nivea*, than a sheep. It thus seems clear that the most useful way of learning about the lethal potential of any snake species to humans is by studying a good cross-section of clinical cases in humans.

Snake venoms serve to incapacitate prey and aid in defence against predators. They might also help to begin and aid digestion, at least in vipers, but this potential additional function is not yet well understood. They consist of various enzymes

ABOVE Collett's snake, *Pseudechis colletti*, an Australian member of the elapid family, characterized by the possession of predominantly neurotoxic venoms.

and other proteins and are among the most complex of all biological toxins. Some of these proteins, such as phospholipase A_2, are particularly widespread enzymes found in the venom of many species, while others are specialities restricted to smaller groups. For example, the venom of stiletto snakes (atractaspids of the genus *Atractaspis*) contains a series of unique amino-acid peptides, named sarafotoxins, that have the specific effect of constricting blood vessels and are not known in other snake venoms. Venom composition varies considerably, not only among species, but sometimes among populations of the same species, and even within the life of an individual. For example, Brazilian lanceheads (the crotaline viperid *Bothrops moojeni*), feed on frogs and lizards as juveniles, whereas adults eat small mammals. This change in dietary habits as they grow larger is accompanied by a corresponding change in venom toxicity to their target prey. Some snakes have subtly different venoms among different individuals of the same species. This is not well understood, but might be linked to differences in diet and/or natural variation across large geographic ranges.

Various animals that prey on venomous snakes have some degree of natural resistance to their venom. Mongooses are well known for their ability to survive bites from cobras, and the tayra, a Central American member of the weasel family, is able to withstand large doses of the venom of the terciopelo (the crotaline viperid *Bothrops asper*). The common mussurana, *Clelia clelia*, a large dipsadine colubrid snake from Central and South America, is immune to snake venoms. The potential medical applications of venom resistance have increasingly become the subject of high-investment research. Potentially beneficial biomedical aspects of snake venoms are also increasingly intensely researched, with multiple products already widely applied in human health. Captotril (capoten), a multi-million-dollar drug developed for treating high blood pressure, for example, is based on a component originally found in the venom of the jararaca, *Bothropoides jararaca*, a South American pit viper. Other venom components are used as diagnostic tools for clotting disorders (RVV-V from *Daboia russelii* and Ecarin from *Echis carinatus*) while others are used as treatments for perioperative bleeding.

There are conflicting ideas among scientists about the evolutionary origins of reptile venoms. It is generally agreed that venom had a single origin within snake evolution (though subsequently reduced or lost in some snake lineages), but there is a lack of clarity about whether venoms in some lizards evolved independently from those in snakes. Whatever its origins, the morphological features adapted to deliver venom into prey are much less sophisticated in the relatively few venomous lizards alive today, such as the gila monsters and beaded lizards (*Heloderma* species). Many of the advanced features of snake venom evolved only after snake evolution was well underway. For example, the so-called three-finger toxins (named for their molecular structure), of which neurotoxins are a member, originated in the Elapoidea + Colubridae lineage after it diverged from vipers. This might have been one of the factors behind that major lineage diversifying into so many species.

Venom is formed from a complex mixture of molecules, and it can be a substantial drain on the resources of an animal. Thus, it is unsurprising that the ability to make potent venom has been lost in many snakes that have evolved different behaviours or diets; for example, in several egg- and snail-eating snakes, and in constricting rat snakes. Molecular biology studies have greatly advanced the understanding of snake venom composition and toxin evolution, but there is still much to learn.

VENOMOUS OR HARMLESS

ABOVE Spix's coral snake, the elapid *Micrurus spixii* (left) and one of its harmless mimics, the South American colubrine colubrid *Simophis rhinostoma* (right).

Among the many remarkable features of snakes, one that attracts much interest is the strikingly similar appearance that some harmless species have to venomous species. This apparent 'mimicry' appears to have evolved primarily to escape predation, and its survival value is clear. The New World coral snakes, for example, advertise their noxious character with a pattern of red, yellow and black rings that say 'Stop! Danger!', and this livery of aposematic (warning) colours is imitated by no fewer than 115 other species – about 18% of all American snakes. One of the most impressive of all mimics is the Harlequin snake, *Pliocercus elapoides*, a dipsadine colubrid from Central America that is known to imitate several different species of elapid coral snake throughout its range and also even their locally specific colour variants (see p.45).

It is not only other snakes that appear to have adopted mimicry as a means of evading predators some marine eels are similar in their banded coloration to the venomous sea kraits (the elapid genus *Laticauda*), and a striking mimic of the eyelash palm pit viper, *Bothriechis schlegelii*, from tropical America exists in the unlikely form of a caterpillar. *Hemeroplanes triptolemus* is a species of hawk moth found throughout much of the eyelash viper's range, and the final stage of its larva bears a resemblance to this pit viper. At rest it looks like a twig, but when alarmed it turns its forebody upside down, withdraws its legs into the body cavity, and expands the thorax into a remarkably lifelike impersonation of an eyelash viper's head (see p.116). The whole transformation is completed in just a few seconds. The third and fourth larval stages of *Hemeroplanes* are said to mimic a different snake, the colubrid *Oxybelis aeneus*.

Reproduction

Most snakes are oviparous, reproducing by laying eggs, although approximately a quarter of all species are viviparous, giving birth directly to young. Viviparity is sometimes commonly called 'live-bearing' but this is not the best term because embryos inside laid eggs are also alive. Sometimes, both kinds of reproduction occur within a single species in different parts of its range. In many snakes, males are more abundant than females, but there are at least two species in which the entire population appears to consist only of females. These snakes reproduce without males by means of parthenogenesis ('virgin birth'), and the young snakes are born as miniature clones of their mothers (see p.58).

COURTSHIP AND COPULATION

In almost all snakes there is a pattern of courtship that precedes mating, and though details vary between different species, in most (and the best-studied) species it tends to involve a similar general sequence of events. The male, which is generally the more active partner, searches out a receptive female by following the scent trail that she produces from special glands as she moves around. Once he has located her, he approaches and proceeds to work his way gradually forwards over her body with rapid, quivering movements, at the same time rubbing his chin along her back and flicking his tongue in and out constantly. In many pythons and boas the male uses the claw-like vestiges of his hindlimbs to scratch or stroke the female's skin, which appears to have the same stimulative effect. When he reaches the nape of her neck, the male then manoeuvres himself into a mating position by throwing a loop of his body over the lower part of her back and entwining his tail around the opening of her cloaca (the common chamber of the reproductive and digestive tract). If the female is receptive, she responds by raising her tail slightly and opening her cloaca, then the male everts one of his paired hemipenes and copulates with her. Although long known to occur in lizards, turtles and crocodilians (as well as mammals), the presence of clitorises

LEFT A pair of smooth snakes, the colubrine colubrid *Coronella austriaca*, mating in southern Britain.

in snakes was only confirmed in 2022. As in lizards, these are paired structures in the base of the tail, termed hemiclitores, though unlike in lizards these are non-eversible structures in snakes. Compared to snake hemipenes, the anatomy, diversity and function of hemiclitores are woefully understudied. Occasionally, intersex snake individuals have been reported, these have fully developed ovaries but also a single hemipenis that is smaller and less-prominently ornamented than the paired hemipenes of males of the respective species.

Mating can be a protracted affair, and the male and female may remain joined together for many hours. The duration of copulation might be determined partly by the morphology of the male hemipenis and the number of males a single female will mate with each breeding season. In species in which females are likely to mate with multiple males, copulation is often longer and the males can insert a secretion immediately after sperm transfer that forms a 'copulatory plug' in the female reproductive tract that reduces the likelihood of insemination by other males.

In some species, the female may be surrounded by considerable numbers of males, which in their impassioned attempts to mate with her and 'squeeze' out competing rivals will often entwine themselves together in a large, jumbled mass. Herpetologists have observed this 'balling' behaviour in various different snakes, especially aquatic species such as anacondas (*Eunectes* species), green water snakes (the natricine colubrid *Nerodia cyclopion*), Arafura Asian and Australasian file snakes (the acrochordid *Acrochordus arafurae*), and some sea snakes. A breeding ball of anacondas may consist of more than 10 males, all coiled around a single female, and they will stay knotted together like this for several weeks. Fertilization may not necessarily happen immediately after mating. Sperm can survive in the female reproductive tract for months, often stored in specialized tubules in the wall of the posterior and (in non-scolecophidians) the anterior parts of the female oviducts (tube-like structures holding eggs when they leave the ovaries). There are instances where the females of some species in captivity have been kept away from males for several years and then unexpectedly given birth to healthy offspring.

Soon after emerging from their winter dens, most males of the North American natricine colubrid, *Thamnophis sirtalis parietalis,* temporarily produce female pheromones that attract other males and switch off male courting behaviour. These transvestite or 'she-male' snakes appear to benefit from this female-mimicking behaviour by conserving energy resources that have been reduced during overwintering. Later in the season the males produce typical male pheromones and use their increased energy levels to begin mating.

RITUALISTIC COMBAT

Should they encounter each other during the breeding season, the males of some species become aggressive and perform ritualistic combat 'dances'. These competitions take place between adult males of the same species, and usually between individuals that are similarly matched in size. When two such rivals meet, they rise up vertically against each other and entwine their bodies like a twisted rope, each attempting to dominate the other by pushing him over. The display is frequently accompanied by exaggerated swaying movements as each snake attempts to knock the other off balance, and finally ends when the weaker opponent concedes by dropping to the ground and moving away.

Ritualistic combat dances have been observed in species as widely different as American indigo snakes (the colubrine colubrid *Drymarchon corais*), European vipers (*Vipera* species), and in Africa black mambas (the elapid *Dendroaspis polylepis*), and Australian tiger snakes (the elapid *Notechis*). Essentially, they amount to little more than a vigorous test of strength, and, except in a few species in which a frustrated male may bite his rival, neither snake usually inflicts serious injury on the other. So preoccupied can a male become in his battle for supremacy that he may become completely oblivious to what is happening around him; a male European adder, *Vipera berus*, for example, will sometimes continue to 'dance' in this way even when its rival has been removed and replaced with a stick.

LEFT Male combat in the western diamondback rattlesnake, the crotaline viperid *Crotalus atrox*, photographed in October in the Tucson Mountains, Arizona, USA.

EGGS, HATCHING, AND BIRTH

Eggs are usually laid several weeks or months after mating, and they vary greatly in number according to species and, in many cases, the size of the parent female. Most of the smaller kinds of snakes, such as scolecophidians and some invertebrate-eating members of the family Colubridae, typically produce only two or three eggs that in comparison with their body sizes are enormous, while

MALE OR FEMALE?

ABOVE Adult female (above) and male (below) individuals of the pseudoxyrhophiid *Langaha madagascariensis* have strikingly different snout processes.

There are often few distinguishing external features between male and female snakes. Sexual dimorphism in these animals is typically limited externally to subtle differences in tail length and perhaps the relative numbers of scales along the belly (ventrals) and underside of the tail (subcaudals). There are, however, some snakes that differ more noticeably. Madagascan leaf-nosed snakes, the pseudoxyrhophiid *Langaha madagascariensis*, for example, have differently shaped appendages on the snout, while males and females of some sea snakes have different colour patterns.

In many snakes, females are larger than males, including natricines, pythons and especially anacondas. One of the more sexually dimorphic snakes studied

to date is the Arafura acrochordid, *Acrochordus arafurae*. Females of this species are conspicuously larger and more heavy-bodied than males, have relatively larger heads and jaws, and shorter tails. An adult female may attain a length of 2 m (6½ ft) and weigh more than 2 kg (4 lb 6 oz), whereas the much smaller males rarely exceed 1 m (3¼ ft) in length and average considerably less than 1 kg (2 lb 3 oz) in weight. When they are about the same length, the female's head is larger and its body much thicker – these differences are apparent even in newborns.

Divergence in body size between the sexes of Arafura acrochordids has probably arisen through evolutionary selection for reproductive success – the larger the female's body, the larger her capacity for producing more offspring in a litter. The difference in relative head and jaw sizes, however, appears to be more the result of adaptations of males and females to different ecological niches rather than sexual selection – the sexes have undergone independent specialization over time to take advantage of different food resources and thereby increase their success rates in hunting. Females hunt in deeper water than males and generally eat only a single large fish, whereas males tend to inhabit shallower water and eat a larger number of smaller fish. As adults, the sexes also feed on different species of fish.

large pythons may lay up to 100 relatively small eggs. Viviparous species in particular often give birth to large numbers of offspring; one puff adder, *Bitis arietans*, was recorded as producing 156 young in a single litter.

A female snake usually deposits her eggs underground or in a shallow hole on the surface covered with leaves, where they are concealed from predators and insulated from fluctuations in temperature and humidity. The incubation temperature and moisture level to which the eggs are exposed have a direct bearing on their development, and it is thus vitally important that she selects her nest site very carefully. A temperature increase of only a few degrees can halve the time required for the eggs to hatch, but if the nest area becomes too hot or does not retain enough heat the developing embryos within the eggs will perish. In a particularly suitable place, females may deposit their eggs communally and even return to the same site year after year. Favourite nesting sites of the grass snake (the natricine colubrid *Natrix helvetica*) in Britain, for example, are garden compost heaps; when these decompose, they generate heat, providing an ideal environment for incubation.

Although most snakes leave nest sites and have nothing further to do with the eggs once they are laid, the females of some species, particularly pythons, some elapids (notably king cobras, *Ophiophagus*) and a number of tropical vipers, remain with their clutch during the entire incubation period. Female pythons are unusual in also being able to control the temperature at which their eggs develop (see p.78). The time that elapses before the eggs hatch varies greatly among species; in some snakes hatching may occur only a few days after the eggs have been laid, while in others it may take more than three months. At hatching time, the young snakes break through the leathery shell using an 'egg-tooth', a small projection on end of their snouts, and this is discarded shortly afterwards. The young of viviparous snakes are born enclosed within a membranous sac, which ruptures soon after birth.

Predator evasion and defence

Snakes show a wide range of behaviour in response to predators, and the defensive repertoires of some species are elaborate. Almost invariably, however, a snake's first reaction when confronted with danger is to try to escape observation, either by remaining motionless or withdrawing out of sight. Most species are coloured and patterned in a way that conceals them in their natural habitat, reducing the risk of detection by predators and helping them stay hidden from potential prey; this is known as procrypsis. The complex geometrical markings of Gaboon vipers, *Bitis gabonica*, for example, are extremely effective in disguising these large African snakes among fallen leaves on the forest floor, while the mottled green patterns of some palm pit vipers (*Bothriechis* species) are perhaps unsurpassed in concealing these species among the greenery of tropical American rainforests. In addition to having camouflage colour patterns, snakes often have a micro-ornamentation on their exposed body scales that absorb or scatter rather than simply reflecting light, so that they do not appear obviously shiny and stand out from the background habitat. A number of tree-living specialists in the family Colubridae are distinctive in having long, thin bodies that resemble vines, and some, such as those of the genus *Oxybelis* and *Ahaetulla*, also have the habit of moving with an irregular swaying action, mimicking a vine or branch trembling in the breeze. Some snakes do not rely on camouflage but instead appear to try to confuse predators. For example, some crotaline viperids hide their heads, thrash,

ABOVE Snakes that rely on immobility to evade detection have complex colour patterns that serve to break up the outline of the body and often resemble dead leaves, moss, bark etc., such as this African Gaboon viper, *Bitis gabonica*.

or raise loops of their body off the ground ('body bridging') when confronted by snake-eating king snakes (species of the North American colubrine colubrid *Lampropeltis*). Several species of neotropical snakes can stiffen the whole of their extended body into a series of short zigzags when disturbed, termed 'body bending', possibly to mimic vegetation, particularly lianas.

WARNING COLORATION

In contrast to colour patterns that provide concealment, those of some snakes draw attention and warn predators of their owners' dangerous capabilities. Aposematic (warning) colour patterns, as they are known, are exhibited also by a number of mostly harmless 'mimic' species, which are assumed to derive protection from them in much the same way (see p.38).

LEFT Neotropical elapid coral snakes, such *Micrurus apiatus* from Central America, have striking colour patterns that appear to be instinctively avoided by many predators. Similar markings are seen also in some harmless 'mimic' species, such as the false coral snake shown above.

BELOW LEFT False coral snakes, the Central American dipsadine colubrid *Pliocercus elapoides*, have long, fragile tails that are easily broken and thus provide the snake with more than one chance of escape. These harmless snakes also have warning colour patterns, similar to those of venomous true coral snakes.

COLOUR CHANGE IN SNAKES

ABOVE AND RIGHT Mussuranas (the dipsadine colubrid *Clelia clelia*) undergo a dramatic colour change from juvenile (above) to adult (right).

Some snakes, including species of *Tropidophis* have the unusual ability to change colour, a curious phenomenon that appears to be associated with activity and generally follows a 24-hour cycle. Among the most striking is that demonstrated in *T. haetianus*, which, when active at night, has a pale yellow ground colour with two rows of dark dorsal markings, but during the day is almost completely black. *Tropidophis feicki* undergoes similar daily colour changes and should this species be subjected to a temperature of less than 17°C (63°F), it adopts a transitional colour phase at any time of the day.

Colour change has also been reported in other snakes, including *Boa constrictor*, Pacific boas (*Candoia bibroni and C. carinata*), the Round Island 'boa' (*Casarea dussumieri*) and the western rattlesnake (*Crotalus viridis*). Most notably, the giant Oenpelli python, *Morelia oenpelliensis*, of northern Australia, changes from a drab brown during the day to a ghostly silver-grey at night. A recently described homalopsid mud snake, *Homalophis gyii*, which when found was a deep brown with a red brown dorsum, was also observed changing colour and became almost white after being placed in a dark bucket. Darkening occurs when melanophores

(pigment-containing cell structures) in the epidermis (outermost skin layer) move closer to the skin's surface; this is probably controlled by hormonal cues.

Colour change in individual snakes may also occur with development. Mussuranas (the dipsadine colubrid *Clelia clelia*), from Central and South America, for example, are bright red with a black head and pale collar as juveniles. At about 60 cm (24 in) in length they change quite abruptly to uniform black or dark bluish-grey. Similarly, some arboreal python and boa species undergo a striking colour change with age, from yellow, orange or brown as juveniles to bright green as adults (see p.95). Researchers do not fully understand why adults and juveniles of these particular snakes are so differently coloured, but it is probably mostly because different aged snakes are adapted to facing different environmental challenges, including predation risk.

Some snakes also undergo seasonal changes in colour, usually in response to reproductive cues or changes in their body condition. After they emerge from hibernation in spring, for example, the general ground colour of male European adders, *Vipera berus*, is a resplendent silver-grey, but they are drabber after the reproductive season has finished.

ESCAPE

Should their initial attempts to avoid detection fail, most snakes will try to make good their escape by fleeing. Many of the long, slender terrestrial species in particular are highly agile and when confronted with danger will disappear with a burst of great speed. A startled pink-tailed forest racer (the neotropical colubrine colubrid *Dendrophidion nuchale*), for example, will dash across the forest floor for 20 m (65 ft) or more before stopping, and these whip-like Central American snakes may also escape by flinging themselves spectacularly off high rocky outcrops. The 'flying' snakes (the colubrine colubrid genus *Chrysopelea*) of South and Southeast Asia have an even more impressive way of avoiding danger. These accomplished climbers will throw themselves off the highest branches and, with their bodies flattened, glide down to the ground or a lower branch.

TAIL BREAKAGE AND POKING

The ability of a species to voluntarily cast off its tail (caudal autotomy) to escape a predator is most commonly associated with lizards, but there are also several snakes in which this behaviour has arisen. Whereas the broken tail of many lizards will grow back, snakes cannot regenerate their tails. The mechanism of tail breakage in snakes is also different from that in lizards; in most lizards the break occurs directly across a single vertebral segment, whereas in snakes it occurs between vertebrae. In some snakes tail loss is limited to a single breakage, with no further breaks occurring after the initial one, whereas the mechanism in others, such as the neotropical neck-banded snake, *Scaphiodontophis annulatus* and the harlequin snake, *Pliocercus elapoides*, of Central America, both colubrine colubrids, appears more specialized.

Instead of sacrificing their tail, other snakes use them to poke into a potential predator that has picked them up. A pointed and hard tail tip increases the effectiveness of this and can make it feel more like a bite. This behaviour is seen in a diverse range of snakes, including typhlopid scolecophidians, uropeltid pipe snakes and atractaspidid stiletto snakes.

INTIMIDATION DISPLAYS

Many snakes resort to hissing when confronted with danger, and various species are known to use other sounds as a means of discouraging unwanted attention. The audible threat displays of rattlesnakes produced by the rapid vibration of their tail rattles are particularly impressive. Neotropical lancehead vipers (*Bothrops* species) do not have rattles, but many of these species and several others are able

ABOVE Should their initial attempts at escape fail, many snakes resort to intimidation tactics when confronted with danger. The startle and threat display of the Asian lined rat snake, the colubrine colubrid *Coelognathus radiatus*, is particularly demonstrative.

RIGHT The eastern bandy-bandy, *Vermicella annulata*, an Australian elapid that raises loops of its body off the ground to deter predators.

LEFT Cobras, like this southern African *Naja annulifera*, are noted for their upright hooded threat and defence posture.

to generate similar sound effects by vibrating their tails among dry leaves. Saw-scaled vipers (*Echis* species) and desert horned vipers (*Cerastes* species) produce a rasping sound by rubbing their coarsely keeled body scales together, while western hook-nosed snakes (the colubrid *Gyalopion canum*), and sonoran coral snakes (the elapid *Micruroides euryxanthus*), are noted for the curious 'popping' sounds they make by drawing air in through the cloaca and expelling it forcibly.

Several snakes make threats with their body posture instead of with sound. The most well-known of these are the cobras, which flatten the upper ribs on their neck to form an impressive 'hood' while raising up to a third of the front of their body off the ground vertically. Additionally, stiletto snakes (species of the atractaspidid genus *Atractaspis*) press their heads to the ground while arching their necks. Some tree-dwelling colubrid snakes, including neotropical bird snakes (*Pseustes* species), the Australian tree snake, *Dendrelaphis punctulatus*, and the African vine/twig snakes (*Thelotornis* species) and the boomslang, *Dispholidus typus*, respond to threats by expanding the neck to appear larger. Many snakes also open their mouths wide when alarmed, and some, such as species of the colubrine colubrids *Leptophis* (Central America), *Philothamnus* (Africa) and *Boiga* (Asia), reveal a strikingly contrasting colour when they gape.

PASSIVE RESISTANCE

When other lines of defence have all been exhausted, some snakes resort to various passive forms of behaviour. Several roll themselves into balls, and some will try to confuse an enemy further by raising and waving their short stumpy tails to imitate a moving head; to add to the illusion, the tails of some species are patterned to resemble the head. Asian sand boas, *Eryx tataricus*, for example, have tail markings that consist of a short, dark, horizontal line and a small spot, resembling the mouth and eye.

Another odd form of passive resistance is that used by, among others, the North American hog-nosed snake (the dipsadine colubrid *Heterodon*). When molested, these snakes turn over onto their backs and, after a few convulsive wriggles, lie still with their mouths open and tongues hanging out, as though dead, making

BELOW Royal or ball pythons, *Python regius*, of Africa are among a number of species that habitually roll themselves up into a tight ball when alarmed.

them less appealing to some predators. If the 'lifeless' snake is turned over onto its belly again, however, it gives the game away by promptly rolling back. Even more remarkable is the autohaemorrhaging behaviour of wood snakes (e.g. the dwarf 'boa' *Tropidophis*, see p.67).

As a further deterrent, some Asian natricine colubrids (*Macropisthodon* species) produce a sticky, bitter-tasting substance from neck glands when provoked, and many other snakes will smear themselves with either obnoxious secretions from musk glands at the base of the tail or the contents of their cloacas. These can be especially foul smelling and the odour often persists for many hours, as anyone who has caught a Eurasian grass snake or North American garter snake will not have failed to notice.

The Asian tiger keelback (the natricine *Rhabdophis tigrinus*), is not only venomous but also poisonous, with the poisons produced in specialized neck glands. These neck-gland toxins are sequestered from their diet which consists of toads. Other, closely related natricines have the same neck glands and also likely sequester toad toxins. Some North American garter snakes (the natricine colubrid *Thamnophis*), feed on toxic newts and sequester their toxins, not in specialized glands but in various body tissues, and these also likely offer defence against predation.

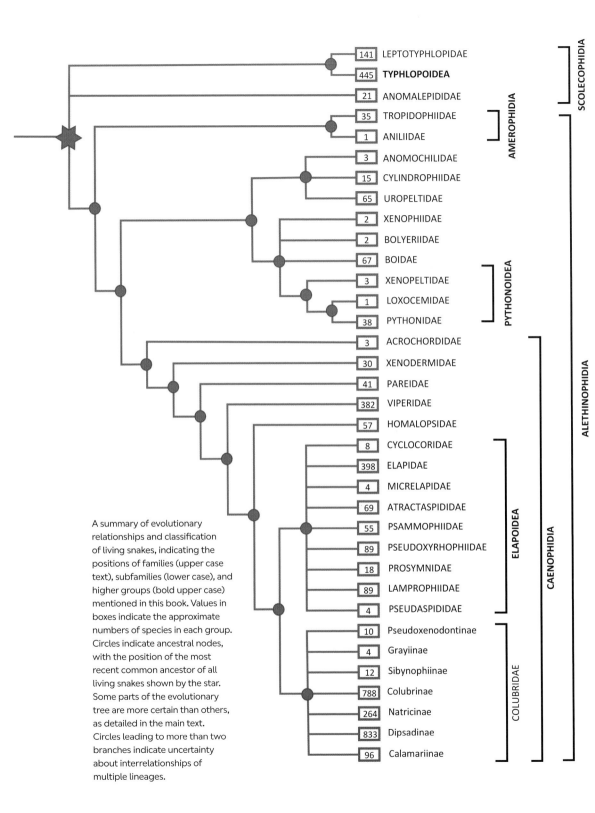

A summary of evolutionary relationships and classification of living snakes, indicating the positions of families (upper case text), subfamilies (lower case), and higher groups (bold upper case) mentioned in this book. Values in boxes indicate the approximate numbers of species in each group. Circles indicate ancestral nodes, with the position of the most recent common ancestor of all living snakes shown by the star. Some parts of the evolutionary tree are more certain than others, as detailed in the main text. Circles leading to more than two branches indicate uncertainty about interrelationships of multiple lineages.

141 LEPTOTYPHLOPIDAE
445 TYPHLOPOIDEA
21 ANOMALEPIDIDAE
35 TROPIDOPHIIDAE
1 ANILIIDAE
3 ANOMOCHILIDAE
15 CYLINDROPHIIDAE
65 UROPELTIDAE
2 XENOPHIIDAE
2 BOLYERIIDAE
67 BOIDAE
3 XENOPELTIDAE
1 LOXOCEMIDAE
38 PYTHONIDAE
3 ACROCHORDIDAE
30 XENODERMIIDAE
41 PAREIDAE
382 VIPERIDAE
57 HOMALOPSIDAE
8 CYCLOCORIDAE
398 ELAPIDAE
4 MICRELAPIDAE
69 ATRACTASPIDIDAE
55 PSAMMOPHIIDAE
89 PSEUDOXYRHOPHIIDAE
18 PROSYMNIDAE
89 LAMPROPHIIDAE
4 PSEUDASPIDIDAE
10 Pseudoxenodontinae
4 Grayiinae
12 Sibynophiinae
788 Colubrinae
264 Natricinae
833 Dipsadinae
96 Calamariinae

SCOLECOPHIDIA
AMEROPHIDIA
PYTHONOIDEA
ALETHINOPHIDIA
ELAPOIDEA
CAENOPHIDIA
COLUBRIDAE

2 The evolutionary tree and classification of snakes

Biological classifications are systems for arranging knowledge about the diversity of life. These filing systems help humans to communicate about organisms, but they can also contain information about how life evolved and is interrelated. During evolutionary diversification, new species can arise when ancestral species split, mostly when populations become separated by different distributions, anatomies, and/or behaviours, to an extent that becomes too great for them to hold together as a single genetic lineage. This mostly divergent, branching nature of evolution has resulted in a hierarchical pattern of relationships among different groups of organisms. Thus, humans are apes but we are also mammals, vertebrates and animals. Most biologists agree that the different ranks in classifications (species, genera, families etc.) are somewhat arbitrary. This arbitrariness becomes less of a concern if we try to make sure that each unit that we name comprises a 'natural group'. A natural group is a complete lineage – one that includes all of the descendants of a particular ancestor. For example, the non-human great apes (gorilla, chimpanzee, orangutan) are not a natural group because their most-recent common ancestor also gave rise to humans. But, the human and non-human great apes together do constitute a natural group (termed the family Hominidae).

Biologists today strive to base classifications on natural groups, and that partly explains why snake classification has undergone several major changes in the twenty-first century, because many discoveries about the evolutionary relationships of snakes have been made in this period. With time and additional data, biologists are likely to agree more on what the natural groups of snakes are, but there will still be debate as to which lineages are assigned the status of genus, subfamily or family, etc. (the arbitrary units above the species level), which explains some of the ongoing changes in classification. Much of this change involves moving species between genera or dividing a single genus into several genera, in order to try to make these groups natural.

We are clearly now in an important age of snake discovery – of new species and of their evolutionary relationships. Undoubtedly the classification used here represents a work-in-progress and it will have changed to some degree by the time the next version of this book is published. The snake classification and framework evolutionary tree used in this book (see p.52) is based on the latest analyses of morphological and DNA data, but new analyses are frequently carried out and new information is continually being discovered. Many species have not yet been studied in enough detail to be included in these evolutionary tree analyses and, on top of this, new species of snake are being discovered all the time and need to be accommodated into the changing classification.

The fossil record, too, will always continue to improve, and it has the potential to overturn some hypotheses based only on studies of living species. Some of the evolutionary relationships among snakes are proving more difficult to resolve than others, so we can expect some further changes to the hypothesized tree over the next few years. For example, evolutionary relationships that are not yet clear or very well supported include whether the Scolecophidia represents one or two major natural lineages; the number of and relationships among major lineages within highly diverse groups such as Colubridae and Elapoidea; and the relationships among Tropidophiidae, pipe snakes and shieldtails (Anillidae, Cylindrophiidae, Anomochilidae, Uropeltidae) and the pythons and boas and their closest relatives. It is also important to note that not all snake biologists will agree on the tree shown on p.52, and even those that do agree with the tree will have different opinions about details of the classification that it supports, in terms of the names and rank used for some of the major natural groups. For example, the group that in this book is referred to as the family Colubridae (comprising at least eight subfamilies) is considered by some snake biologists to instead be a superfamily (termed Colubroidea) with eight families, one of which (Colubridae) is equivalent to the subfamily Colubrinae referred to in this book. There is no one person or committee that decides which of these alternatives is 'correct'. Instead,

RIGHT Snake species are frequently being discovered and named by scientists. The peninsular Indian pareid, *Xylophis mosaicus,* was added to the inventory as recently as 2020.

the community of researchers typically gravitates towards one classification over time, with the ongoing modifications hopefully becoming less frequent and less drastic. Ongoing change and some alternative names for the same groups make some aspects of snake classification confusing and complicated. We cannot here propose a problem-free solution or guarantee long-term stability, but in this book we try to be clear about which names we are using in which way.

On the framework evolutionary tree shown on p.52, the different groups of living snakes are all shown at the tips of the branches of the tree on the right. Each of the branching points (shown by dots) represents the splitting or diverging of the most-recent common ancestor of those living-snake lineages that connect directly to the branching point. The further away from the tips of the tree you look, the further back in time these ancestors would have lived. For example, the most recent common ancestor of vipers and pythons lived further back in time than the most recent common ancestor of vipers and elapids. It is important to realize that the position of a group or particular species on the tree does not mean that it can be labeled as 'primitive' or 'advanced'. Members of the group Caenophidia are sometimes referred to as the 'advanced snakes', but this is a misleading name because all organisms are made up of a mixture of primitive and advanced features. So, although scolecophidians lack fangs and venom, they should not be regarded as 'primitive' snakes. Yes, they might retain some primitive features that have been lost in, for example, vipers (such as remnants of the pelvic girdle), but scolecophidians also have many advanced features adapted for their specialized burrowing and feeding behaviours that are not found in vipers.

LEFT The elapoid *Buhoma procterae* from Tanzania's Uluguru Mountains. The three species of the genus *Buhoma* are more closely related to Elapidae than to Colubridae, but they are not yet assigned to a particular family within Elapoidea because their precise evolutionary relationships remain unclear.

3 Snake diversity

SCOLECOPHIDIA: Worm, Blind and Thread Snakes

Scolecophidia comprises 465 currently recognized, living species, forming two main evolutionary lineages, (1) the family Anomalepididae, and (2) the superfamily Typhlopoidea (itself containing three families) plus the family Leptotyphlopidae. It has been a matter of much disagreement as to whether these two major lineages are each other's closest living relatives (i.e. whether Scolecophidia is a single, natural group) or whether one of the two main scolecophidian lineages is instead more closely related to all other living snakes (Alethinophidia). The latest analyses of extensive DNA data, support the latter hypothesis, and indicate that Anomalepididae is more closely related to Alethinophidia. It will be interesting to see if additional analyses of even more data continue to support this conclusion. Whichever resolution is best-supported, an examination of the evolutionary tree of living snakes (see p.52) shows that Scolecophidia (or one of the two major lineages within it) comprise one half of the primary bifurcation in the tree. Thus, this is an ancient group of snakes, as old as all the lineage including all other living snakes together (Alethinophidia). However, scolecophidians comprise far fewer living species (465) than alethinophidians (approximately 3,600 species) and are also less morphologically (at least superficially) and ecologically diverse.

All the species in the five scolecophidian families are cylindrical, generally small and narrow, have evenly sized body scales, very short tails, and are burrowers in mostly tropical and subtropical sands and soils. They have small, ventrally positioned mouths, very small eyes that are often not visible externally, and they feed mostly on small social insects (ants and termites) and especially their eggs, larvae and pupae. So different is their appearance it is difficult to appreciate that they belong to the same group of vertebrates as the huge pythons, and it was once even questioned whether scolecophidians are truly snakes.

OPPOSITE Western diamondback rattlesnake, *Crotalus atrox*, a crotaline viperid of arid habitats in southwestern USA and Mexico.

Although many scolecophidian species are commonly called 'blind snakes', there is no evidence that they completely lack visual function. Admittedly their eyesight is unlikely to be good. Each eye is not covered by its own, circular protective cap (the brille), as in most other snakes, but usually lies beneath unspecialized head scales. The seven-striped thread snake (the leptotyphlopid *Siagonodon septemstriatus*) from northern South America has a distinct pupil and coloured iris, and may have better developed vision than most other scolecophidians. In many scolecophidians the eyes have been reduced substantially and, although they retain a structured retina and lens, they also lie beneath head scales and might only discriminate light from dark instead of being capable of forming a detailed image.

The natural history of all but a handful of scolecophidian species is little-studied and extremely poorly known. All scolecophidian snakes are probably egg layers, and except for some anomalepidids they have only one oviduct, perhaps an evolutionary consequence of their very slender bodies. Clutch size ranges from a single, long and very narrow – 25 x 2.5 mm (1 x ¹⁄₁₀ in) – egg in the Arabian and Asian leptotyphlopid *Myriopholis blanfordi*, to as many as 60, each about 20 x 10 mm (⁴⁄₅ x ²⁄₅ in), in the giant Schlegel's blind snake (the African typhlopid *Afrotyphlops schlegelii*). The eggs laid by some species are so tiny that in size and shape they resemble a grain of rice. Some species lay eggs that are in a very advanced stage of development that hatch within a few days.

Among the most unusual of the scolecophidian snakes in its reproductive biology is the Brahminy blind snake, the typhlopid *Indotyphlops braminus*, which is an all-female, parthenogenetic species (i.e. it produces offspring without males)

LEFT *Afrotyphlops angolensis*, a large blind snake found throughout much of tropical Africa.

that arose through a natural hybridisation event. This species is also called the 'flower-pot snake' because it is small (rarely longer than 10 cm; 4 in), hides in soil and has been accidentally transported by humans to many places, possibly hidden in containers of crop and ornamental plants. A native of southern Asia, it has now become established in regions as far apart as Madagascar, Japan, Australia, Hawaii, Mexico and the USA.

FAMILY ANOMALEPIDIDAE

The Anomalepididae includes only 21 species in four genera (*Anomalepis*, *Helminthophis*, *Liotyphlops*, *Typhlophis*), restricted in distribution to tropical Central and South America. In evolutionary terms, these snakes seem to have lost some of the 'primitive' features that other scoleocophidians retain, such as vestiges of a pelvis and several other features of the skeleton. Anomalepidids differ from the Leptotyphlopidae and Typhlopoidea in having teeth on both the lower and upper jaws, and in having more rows of scales around the body. Most species are small, with only a few reaching total lengths of over 30 cm (12 in). Very little is known about their natural history, but the few observations that have been made indicate that they have similar diets and habits to most other scoleocophidians. The family Anomalepididae does not have any widely used common English names.

ABOVE *Typhlophis squamosus*, an anomalepidid from northern South America.

FAMILY LEPTOTYPHLOPIDAE: Thread or Worm Snakes

This family of about 140 species includes the smallest and the most elongated species of all living snakes. Leptotyphlopids occur primarily in South America and Africa but extend also into the Caribbean and Central and southern North America, Arabia and southwestern Asia. They have been recorded from an exceptionally large range of elevations, from 76 m (250 ft) below sea level (*Rena humilis* in Death Valley, USA) to more than 3,250 m (10,660 ft) above sea level (*Epictia tricolor* in the Andes of Peru). Previously, only two genera were recognized (divided into a somewhat confusing assemblage of species groups), but recent studies of DNA have helped scientists to tease apart some of the diversity within this old family, and to demarcate 14 genera, the most species-rich of which is *Epictia* from Central and South America (44 species). Although some leptotyphlopids reach over 45 cm (18 in) in total length, *Tetracheilostoma carlae* from Barbados is the smallest known snake with a maximum length of only a little over 10 cm (4 in). Some leptotyphlopids are extremely elongate, with species of the African genera *Myriopholis*, *Namibiana* and *Rhinoleptus* having body widths that are less than 140th of their total length.

In leptotyphlopids the upper jaws have no teeth and are relatively immobile, while the lower (toothed) jaws are highly flexible and are swung backwards and forwards rapidly to rake in ant and termite eggs and larvae. They eat large

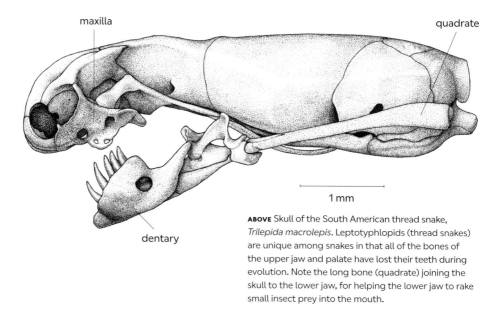

maxilla

quadrate

1 mm

dentary

ABOVE Skull of the South American thread snake, *Trilepida macrolepis*. Leptotyphlopids (thread snakes) are unique among snakes in that all of the bones of the upper jaw and palate have lost their teeth during evolution. Note the long bone (quadrate) joining the skull to the lower jaw, for helping the lower jaw to rake small insect prey into the mouth.

numbers of these small prey items in each meal, with this special rapid feeding mechanism, probably being an adaptation to help them escape from ant and termite nests before suffering attacks from their soldiers. Texas thread snakes, *Rena dulcis*, and perhaps other species too, are thought to form a physical barrier against the bites of these insects by raising the tips of their body scales. Many scolecophidians (and other snakes) produce distasteful secretions that might also help to deter attacking insects.

THREAD SNAKES AND SCREECH OWLS

ABOVE Texas thread snake, the leptotyphlopid *Rena dulcis*.

Eastern screech owls, *Megascops asio*, feed mainly on insects, but among the various other small animals that parent birds in southern USA occasionally catch and take back to their nestlings are Texas thread snakes, *Rena dulcis*. Unusually, however, the owls do not always kill the snakes, as they do most other kinds of prey, but carry them back to the nest alive, where apparently many are then released (or escape from the bird's bill and talons) and survive by eating the larvae of parasitic insects. Some captured snakes continue to live in the nest, feeding on insect larvae even after the young owls have fledged.

Infestations of scavenging and parasitic insects often plague the nests of screech owls, and this appears to be one of the main underlying reasons why in parts of their range their broods so often fail. Young owls in nests that contain thread snakes, however, appear to grow faster and suffer lower mortality than do those in which there are no snakes, so evidently there is at least some benefit to the owls in having these live-in cleaners around. It remains to be seen if the occurrence of thread snakes in screech owl nests is merely the result of them having fortuitously escaped being eaten, or if there is some complex, mutually beneficial interaction between these animals at play. Where they occur in other parts of the world, thread snakes are often preyed on by owls and invariably eaten by these birds.

SUPERFAMILY TYPHLOPOIDEA: Blind Snakes

Typhlopoidea comprises three families: 280 species of Typhlopidae (the largest family of scolecophidians) almost 30 species of Gerrhopilidae, and a single species of Xenotyphlopidae. The single xenotyphlopid species, *Xenotyphlops grandidieri*, occurs only on Madagascar where it burrows in loose, sandy coastal soils. Gerrhopilidae currently contains two genera: the mysterious *Cathetorhinus* (which some biologists are not even sure is a gerrhopilid) is known only from a single specimen collected between 1801 and 1803 from an unknown locality, and 28 species of *Gerrhopilus* distributed from India and Sri Lanka to western Melanesia.

The 280 species (in 18 genera) of Typhlopidae are geographically widespread, occurring in the Americas, Africa, Europe, Asia, Madagascar and Australia. An African typhlopid species, Schlegel's blind snake, *Afrotyphlops schlegelii*, is the largest known scolecophidian, growing to almost 1 m (3¼ ft) long and about 3 cm (1⅕ in) in diameter. There are several other 'giant' typhlopids that reach lengths of more than 60 cm (24 in), but in general the length of these snakes is 15–30 cm (6–12 in). All species, including the largest, seem to feed exclusively on small invertebrates. Typhlopids are generally more stocky than leptotyphlopids, but there are a wide range of body shapes. The arboreal blind snake, *Ramphotyphlops*

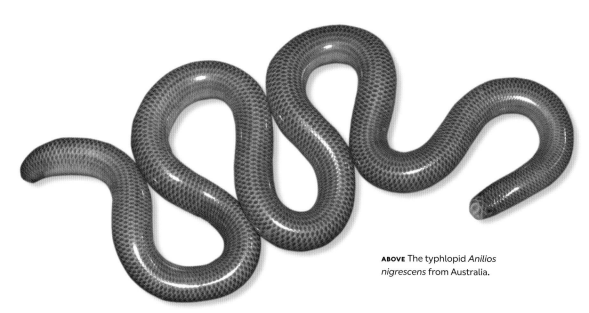

ABOVE The typhlopid *Anilios nigrescens* from Australia.

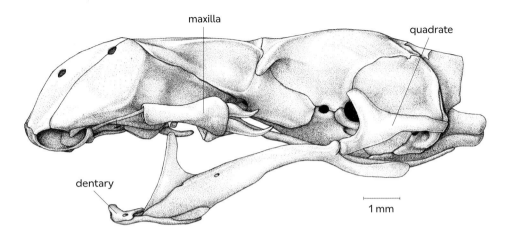

ABOVE Skull of *Afrotyphlops punctatus*, an African 'blind' snake (family Typhlopidae). The upper jaw has a couple of teeth on each maxilla but the lower is toothless. Unlike leptotyphlopids the maxillary bones of the upper jaw are also not rigidly attached to each other but capable of some independent movement.

angusticeps, from the Solomon Islands has a particularly long body with more than 600 individual vertebrae, perhaps the greatest number of any snake (or any other vertebrate). Several typhlopids seem to lack body pigment entirely, as do some xenotyphlopids and gerrhopilids. In life these species are bright pink thanks to the oxygenated blood in their translucent bodies.

Superficially the heads of typhlopoids are like those of leptotyphlopids, but internally there are surprising differences. In almost a complete reverse of the leptotyphlopid condition, the upper jaws have teeth and are freely moving while the lower jaws are toothless and rigidly attached to each other anteriorly. Like the leptotyphlopids they can feed very rapidly (swallowing nearly 100 small prey items in one minute) by raking, but in typhlopoids it is the upper jaws that perform this function. The typhlopid Bismarck sharp-nosed blindsnake, *Acutotyphlops subocularis*, of the Bismarck Archipelago off the northeastern coast of New Guinea are unusual in that they feed mostly on earthworms, which are swallowed whole.

AMEROPHIDIA: Dwarf 'Boas' and Neotropical Pipe Snakes

Amerophidia forms one half of the oldest evolutionary divergence among alethinophidian snakes, the other branch being Afrophidia. Amerophidia comprises approximately 100 times fewer living species than Afrophidia, and so must have undergone much less evolutionary diversification and/or greater levels of extinction since the two branches diverged in the Cretaceous Period. The two major lineages of Amerophidia are highly disparate, the burrowing neotropical pipe snakes (Aniliidae) and ground-dwelling and arboreal dwarf 'boas' (Tropidophiidae). Although snake biologists were not entirely surprised when early DNA analyses began to indicate that tropidophiids were not especially closely related to true boas (and particularly to the banana and Oaxacan boas, see p.92), many of them were puzzled by evidence indicating that tropidophiids are most closely related to neotropical pipe snakes because there is little in their anatomy to suggest this. Subsequently, additional DNA evidence has further strengthened this conclusion, though supporting evidence from anatomy remains scarce, possibly because the two groups have very different lifestyles.

FAMILY ANILIIDAE: Neotropical Pipe Snakes

Aniliidae comprises a single species, *Anilius scytale*, found in the wet forests of the Amazonian and Guiana Shield regions of South America. This species is sometimes called the false coral snake because of its overall body shape (cylindrical with a short tail and narrow head), partly burrowing lifestyle, and exceptionally bold black-and-red colour pattern. Unlike true coral snakes (family Elapidae), *Anilius* is very mellow and non-venomous. It relies on its warning coloration and body-flattening defensive behaviour to keep predators at a safe distance. It has very small eyes that lie beneath a relatively large, transparent, polygonal scale. This species retains the primitive feature for snakes of teeth on the premaxilla. Although mostly burrowing in habit, *A. scytale* can be found in and near water, and many individuals have been found moving above ground at night. It gives birth to 3–13 young. The narrow-mouthed *Anilius* seems to feed only on long-bodied and narrow prey, including other snakes, caecilians (burrowing, limbless amphibians), amphisbaenians (burrowing limbless lizards) and eels.

ABOVE The strikingly marked South American pipe snake, the aniliid *Anilius scytale*, grows up to about 90 cm (3 ft) long.

FAMILY TROPIDOPHIIDAE: Dwarf and Eyelash 'Boas'

In certain features of anatomy, the two genera of small boa-like forms in the family Tropidophiidae resemble the larger, true boas of the family Boidae, but they are only distant relatives. With the exception of the yellow banded dwarf 'boa', *Tropidophis semicinctus*, they share the primitive feature of having a pelvic girdle and, in males of most species, external vestiges of hindlimbs. Tropidophiids have a well-developed tracheal lung, and the left lung is greatly reduced or lacking altogether. Neotropical in distribution, tropidophiids are small snakes with bodies that are cylindrical or slightly flattened from side to side, and most have relatively short, prehensile tails. Generally secretive and essentially nocturnal inhabitants of the forest floor, several *Tropidophis* are at least partially arboreal. Many are active foragers, feeding mainly on lizards and frogs, which they kill by constriction. The larger species of *Tropidophis* may occasionally also feed on small rodents and nestling birds. Except for the southern eyelash 'boa', *Trachyboa gularis*, all species are viviparous.

ABOVE At slightly over 1 m (3 ft) long, the tropidophiid *Tropidophis melanurus*, from Cuba, is among the largest members of its family. This orange-coloured example is one of two main colour forms.

Prominent among the snakes of this family are 33 species of *Tropidophis*, most of which are distributed among the islands of the Caribbean, where the scarcity of other snakes has perhaps permitted these forms to diversify and exploit a wide range of habitats. The genus is best represented on Cuba, with some 17 species found there. Five species survive as relicts on the South American mainland, ranging from Ecuador (*T. battersbyi*) to Peru and Brazil (*T. paucisquamis*, and *T. taczanowskyi*, *T. preciosus* and *T. grapiuna*).

Most *Tropidophis* are dull brown or grey snakes with a variable pattern of dorsal body blotches and spots. Three Cuban species, *T. semicinctus*, *T. spiritus* and *T. wrighti*, are more conspicuously marked, with patterns of pale and dark bands or blotches. The scales on the dorsum are keeled in some species and smooth in others. Sometimes also called 'wood snakes', *Tropidophis* occur predominantly in forested habitats, ranging from humid rainforest to dry Acacia–cactus scrub, and one relatively recently described Cuban species (*T. fuscus*) is known only from pine woods associated with red lateritic soils in the eastern part of the island. Most *Tropidophis* are terrestrial and typically found beneath logs, fallen palm fronds and other forest floor debris, and in ant and termite nests. Some species are more arboreal, often feeding on frogs. At least some species likely use caudal luring (see p. 29) to attract prey.

Tropidophis are unique among snakes for their extraordinary ability to autohaemorrhage, a form of behaviour that appears to have evolved as a form of defence. If disturbed, their eyes turn red with blood, and the mouth begins to bleed freely from veins on the palate. Although alarming to observe, this does not seem to have any detrimental effect on the snake and, when danger has passed, it will quickly recover. It is not clear whether *Tropidophis* blood is actually an irritant (or poison) or whether the bleeding simply puts potential predators off. If molested, a *Tropidophis* may also coil up in a small ball with its head hidden in the centre, and produce an offensive-smelling anal secretion.

Various other snakes (all of them North American colubrids) are reputed to adopt 'bleeding' behaviour. Long-nosed snakes (the colubrine *Rhinocheilus lecontei*) and eastern hognosed snakes (the dipsadine *Heterodon platyrhinos*) bleed from the cloaca, while the yellow-bellied water snake (the natricine *Nerodia erythrogaster*) sometimes exudes blood from the gums. In these species the bleeding may be incidental, caused by the wild thrashing they often resort to when molested, but in *Tropidophis* it appears to be more controlled and deliberate.

ABOVE The northern eyelash 'boa', *Trachyboa boulengeri*, a tropidophiid from lowland forests of northwestern South America.

The other genus in the Tropidophiidae, *Trachyboa*, contains two small, stout-bodied, somewhat arboreal species with an extremely short tail and strongly keeled scales. The northern eyelash 'boa', *Trachyboa boulengeri*, has one or more enlarged, projecting scales over the eye and similar horn-like scales on the canthus (contour of the snout between the top and side of the head), a feature that has given rise to the name 'eyelash boa', by which both species are known. They feed almost exclusively on small frogs. *Trachyboa boulengeri* occurs in the humid lowland rainforests of Costa Rica, Panama, Ecuador and Colombia, while the other species, *T. gularis*, is found in the dry coastal forests of western Ecuador.

CYLINDROPHIIDAE, ANOMOCHILIDAE AND UROPELTIDAE: Asian Pipe Snakes and Shieldtails

FAMILIES CYLINDROPHIIDAE AND ANOMOCHILIDAE: Asian Pipe Snakes

Cylindrophiidae is represented by 15 species classified in the genus *Cylindrophis*, found across Indochina and the islands of Southeast Asia, with a single species (*C. maculatus*) in Sri Lanka. The single genus (*Anomochilus*) and three species in the family Anomochilidae are restricted to peninsular Malaysia and Borneo. Asian pipe snakes were previously thought to be closely related to the superficially similar neotropical pipe snakes (Aniliidae), but their superficial similarity is now known to be better explained by adaptation to a similar lifestyle of burrowing in moist tropical soils and feeding mostly on narrow, elongate prey, rather than by a close evolutionary relationship.

RIGHT The anomochilid *Anomochilus monticola* from the Mount Kinabalu region of Borneo. Note the slender, cylindrical body and iridescent scales.

ABOVE The Sri Lankan pipe snake, *Cylindrophis maculatus*, attempts to distract potential predators by flattening its body, hiding its head, and curling its tail conspicuously.

When danger threatens, cylindrophiids can flatten and curl their tails upwards to expose boldly patterned and coloured undersurfaces. The dorsal colour of some species is a uniform brown, while others are marked with black cross bars and a lengthways stripe along the middle of the back, and most species have black-and-white-chequered bellies. Like other pipe snakes, they are burrowers in loose, wet soil or leaf litter, usually close to the surface. They live in rainforest and wetlands, but some species can also be found in agricultural habitats. As far as is known, they are all viviparous. *Cylindrophis* have been recorded feeding on eels and caecilian amphibians but have also been known to accept small fish in captivity.

Anomochilids are small snakes, growing to only slightly over 50 cm (20 in). They are marked with pale spots, and two species have bright red bands on the tail. Unlike other pipe snakes, they are egg-layers, although apart from this (and that they live in rainforests) very little is known about their natural history. The diets of *Anomochilus* are unknown, but the small head and body size of these little snakes indicate that they probably eat only narrow prey such as small caecilians and snakes or perhaps worms and other invertebrates.

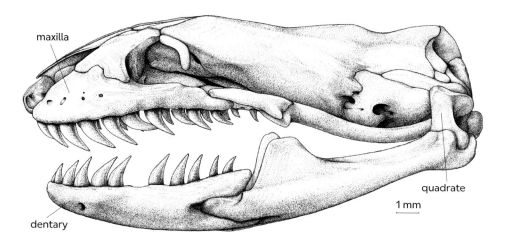

maxilla

quadrate

1 mm

dentary

ABOVE Skull of the cylindrophiid pipe snake, *Cylindrophis ruffus*. Pipe snake skulls retain several primitive features for snakes and have only a few large teeth. The snout and jaw bones are not very movably connected to each other or the braincase, as in most other snakes, and the lower jaw, too, is relatively inflexible so that these snakes do not eat wide prey.

FAMILY UROPELTIDAE: Shieldtails

The shieldtails, of which there are seven genera (*Melanophidium*, *Platyplectrurus*, *Plectrurus*, *Pseudoplectrurus*, *Rhinophis*, *Teretrurus*, *Uropeltis*) and approximately 65 species, are a fascinating group of distinctive-looking snakes restricted to Sri Lanka and the Indian peninsula. They spend most of their time burrowing in moist soils, particularly in mountainous regions. They are most closely related to Asian pipe snakes, and they resemble them in being cylindrical, feeding on narrow prey and having a robust skull that retains several primitive features. They differ from pipe snakes in having no or fewer vestiges of the hindlimbs and pelvic girdles, and most are also even more radically modified for subterranean life.

The most notable feature of shieldtails, and the feature for which they are collectively named, is the oddly shaped end of the tail. In some species this is capped with a spine or has an enlarged, roughened scale, while in others the tail terminates abruptly as a broad, flattened disc covered with many spines or sharp keels. Internally the tail shield is supported by a bony plate. The curiously modified tails of these snakes probably serve some adaptive function in protecting them; when threatened with attack, a shieldtail will tuck its head between or under body coils and wave its tail around to make it look more like a head, a behavioural trait that probably evolved as a means of diverting a predator's attention away from the snake's more vulnerable head.

LEFT The Indian uropeltid *Uropeltis macrolepis mahableshwarensis*. Note the flattened shield-like form of the tail.

Shieldtails are small snakes, 20–75 cm (8–30 in) in length, with most species towards the lower end of this size range. Some species are brilliantly coloured, with bold red, orange and/or yellow markings, while others are uniformly black. The bright colours are thought to help distract predators, such as jungle fowl, that dig around in leaf litter and shallow soil looking for prey. As with many other groups of burrowing snakes, the scales of uropeltids are often highly iridescent. They occur mostly in mountain rainforests and agricultural plots that have replaced this native habitat, although some, such as *Rhinophis dorsimaculatus* of Sri Lanka, occur in sandier soils of coastal areas. They feed mainly on earthworms, and some species can be found in high densities. Although living mostly within the soil, shieldtails can be seen on the surface during and shortly after heavy rains, perhaps to avoid drowning in burrows, but probably also to feed on surface-active earthworms.

BOLYERIIDAE AND XENOPHIIDAE: Round Island 'Boas' and Spine-jawed Snakes

The 'boas' of Round Island (family Bolyeriidae) are of particular interest to biologists because of their uncertain origin and isolated distribution. Until recently, two species were known from this small landmass in the Indian Ocean, although owing to large-scale habitat destruction by goats and rabbits introduced in the nineteenth century, one (*Bolyeria multocarinata*) is probably now extinct. A single individual captured in the mid-1970s was the last to have been seen alive. Subfossil remains of a third, extinct species have been found on the nearby island of Mauritius.

ABOVE The Round Island boa, *Casarea dussumieri*, is likely the only surviving species in its family, the Bolyeriidae.

Round Island 'boas' were previously classified as a subfamily of the true boa family (Boidae). Unlike that of boids, however, the left lung is relatively much shorter (only about 10% or less of the length of the right lung), and there is no pelvis or vestiges of hindlimbs. In respect of these and certain other features, Round Island 'boas' resemble the 'higher snakes' (Caenophidia). *Bolyeria multocarinata* appears to have been a burrowing form, whereas the surviving species, *Casarea dussumieri*, is ground-dwelling and reaches a length of approximately 1.2 m (4 ft). Nocturnal in habit, *C. dussumieri* has been found by day hiding beneath fallen palm fronds, in the lower branches of trees, and in burrows excavated by nesting seabirds. It is oviparous, and feeds almost exclusively on the island's few endemic species of geckos and skinks. Although pushed to the brink of extinction, it is the target of conservation efforts, and its numbers are seemingly increasing.

The family Xenophiidae comprises only two species in the genus *Xenophidion* (meaning 'strange snake'). Both species were described as recently as 1995 (although they were discovered nearly 10 years prior to this) and they are known only from a total of five individuals. They have scale patterns and jaw anatomies that are quite divergent from any other known group and, as with the Round Island 'boas', researchers have struggled to interpret the family's relationship to other snake families. The two families might not be especially closely related, but they are included in the same section here as two somewhat mysterious groups that are possibly closely related to pythons and sunbeam snakes. The 'common' name for Xenophiidae, spine-jawed snakes, comes from a spiny projection that extends backwards from the upper jaw in both species. *Xenophidion acanthognathus* occurs in Borneo, and *X. schaeferi* in peninsular Malaysia, both in rainforest areas. Due to the scarcity of specimens and data, almost nothing is known of their natural history. They are likely secretive and possibly burrowing. Examination of museum specimens indicates oviparity and a diet that includes at least skinks.

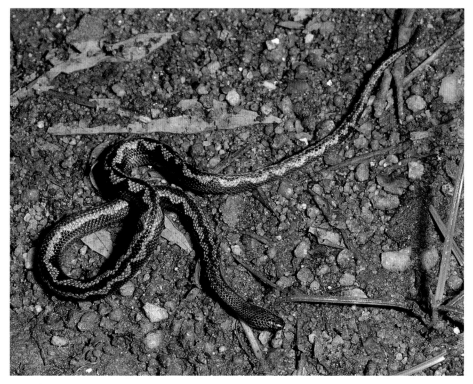

ABOVE From peninsular Malaysia, the very rarely encountered *Xenophidion schaeferi*.

ABOVE The Antiguan racer, *Alsophis antiguae*.

Despite the many threats to snake populations worldwide, conservation efforts are underway to help boost populations of some species. Two such projects that have seen real success both involve island snakes that were brought to the brink of extinction by invasive mammal populations. The Antiguan racer (the dipsadine colubrid *Alsophis antiguae*) became extinct on the Caribbean island of Antigua, and on neighbouring Great Bird Island the population was reduced to only 50 individuals in 1995. Extinction on Antigua was caused by mongooses introduced to control the snake populations, and population decline on Great Bird Island was caused by invasive black rats. By removing black rats from Great Bird and 11 other islands, as well as starting a captive-breeding programme, numbers of what was once considered the 'world's rarest snake' increased ten-fold to 500 in 2010. Similarly, the keel-scaled boa (*Casarea dussumieri*) on Round Island, Mauritius, had been reduced to only 60 individuals in 1976 by invasive rats, and through overgrazing by introduced goats and rabbits that degraded

the snake's habitat. Thanks to a captive-breeding effort, the removal of these non-native mammals, and the reintroduction of native plant species the population of Round Island boas increased by 2,600% by 2012.

Several projects are trying to reverse snake declines and expand population ranges using a mixture of habitat restoration and captive breeding. These include three colubrids, the eastern indigo snake (*Drymarchon couperi*) and the Louisiana pine snake (*Pituophis ruthveni*) in the USA, and the smooth snake (*Coronella austriaca*) in the UK, and the Woma python (*Aspidites ramsayi*) in Australia. However, it is often difficult to prove that snake numbers have increased significantly thanks to captive breeding and reintroduction alone, and in most cases habitat destruction continues to be a huge concern for many snake species. Although many zoos and research institutions are studying and refining techniques for captive breeding and release, few practical examples exist of indisputable, large-scale success, and there is little hard data available as yet to suggest that other conservation projects have resulted in a significant increase in snake numbers.

BOIDAE AND PYTHONOIDEA: Boas, Pythons and Relatives

Although they resemble each other in many features, especially in their adaptations to similar environmental conditions, researchers now accept that boas and pythons are not quite as closely related as was once believed and belong to different families: pythons in the Pythonidae and boas in the Boidae. These are not each other's closest living relatives because Pythonidae is more closely related to the Mexican burrowing 'pythons' and sunbeam snakes (Loxocemidae and Xenopeltidae; pp.76–77) than to Boidae. Thus, pythonids, loxocemids and xenopeltids together form a larger natural group, Pythonoidea. Primitive features for snakes that are found in all boas and pythons are a long row of palatal teeth and a pelvis with vestigial (remnant) hindlimbs, visible as small claw-like spurs at the base of the tail. These spurs are more prominent in males and likely function to stimulate females during copulation. Most species have a relatively large left lung (typically 30–60% as long as the right lung), and some also have supraorbital bones, premaxillary teeth, and small infrared-sensing pits in, behind or between scales along the lips. The bones of the skull are more loosely articulated than in the mostly burrowing snakes covered in preceding sections, enabling boas and pythons to swallow particularly large prey. Anyone who has observed these snakes feeding cannot fail to be impressed by their extraordinary ability to extend the jaws around what often seems to be an impossibly large mouthful.

BELOW The heat sensory organs of boas most typically occur between the labial (lip) scales as exhibited by the emerald tree boa (left). Those of pythons, such as this amethystine python, *Simalia amethistina*, of Indonesia, New Guinea and northern Australia (right), are generally placed within the labial scales.

Boas and pythons are mostly mutually exclusive in their geographic distribution, so there are very few regions where both groups can be found. There are no boas in Southeast Asia or Australia, only pythons. In the Americas and Madagascar, there are only boas. Only in the New Guinea region and parts of India and Africa do ranges of some boa and python species overlap, and in these areas the two groups tend to occupy different habitats. Pythons are distinguished from boas in the arrangement of particular bones of the skull, the placement of their infrared-detecting pits, and teeth on the premaxilla which boas lack. The exception to this is the two species in the Australian python genus *Aspidites* which, like boas, lack premaxillary teeth.

FAMILIES LOXOCEMIDAE AND XENOPELTIDAE: Mexican Burrowing 'Pythons' and Sunbeam Snakes

Loxocemidae and Xenopeltidae are very small families that, although widely separated geographically, are superficially similar and most closely related to pythons. Their scales have an unusually iridescent quality, particularly those of the Asian *Xenopeltis unicolor*, and for this reason they are often referred to as 'sunbeam snakes'. They are semi-burrowing snakes of moderate size – up to approximately 1 m (3¼ ft) in length – and are all egg-layers. The Mexican burrowing 'python', *Loxocemus bicolor*, is the sole living representative of the

RIGHT Neotropical sunbeam snake, *Loxocemus bicolor*.

Loxocemidae. It has vestiges of a pelvic girdle, and also the primitive features of supraorbital bones and premaxillary teeth. *Loxocemus* is a constrictor and feeds on lizards and small rodents, mostly underground. Turtle and iguana eggs also feature in its diet. Originally described from El Salvador, it occurs also in Mexico and the Pacific lowlands of Central America. Despite the common name, *L. bicolor*, is not a true python (Pythonidae).

The Xenopeltidae is an Old World family with one genus, *Xenopeltis*, and three species, all in Asia. Although separated by millions of years of evolution, among living snakes Xenopeltidae is probably most closely related to Loxocemidae plus Pythonidae. In general appearance they resemble *Loxocemus*, especially in body proportions and the appearance of their scales, and similarly they have a relatively long left lung (approximately half as long as the right lung), premaxillary teeth, and a jaw mechanism that is more flexible than that of pipe snakes. Unlike *Loxocemus*, however, *Xenopeltis* lack supraorbital bones or pelvic girdle remnants. Sunbeam snakes have a rather cylindrical body, and the head is somewhat flattened with quite small eyes. They are largely soil- or compost-dwelling and/or semi-aquatic in habit. They feed primarily on snakes, but will also eat small mammals and frogs in captivity. *Xenopeltis unicolor*, the most wide-ranging species, lives in India (Andaman and Nicobar islands), southern China, and throughout much of Southeast Asia, including many of the Indonesian and Philippine islands. Its less well-known relatives, *X. hainanensis* and *X. intermedius*, are known only from China and Vietnam, respectively.

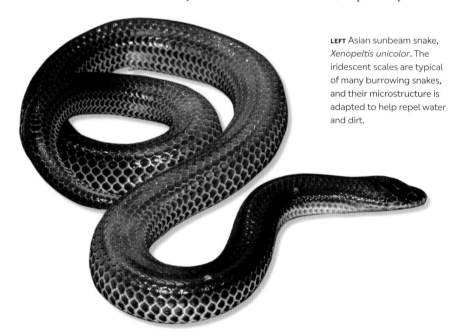

LEFT Asian sunbeam snake, *Xenopeltis unicolor*. The iridescent scales are typical of many burrowing snakes, and their microstructure is adapted to help repel water and dirt.

FAMILY PYTHONIDAE: Pythons

All 38 python species are oviparous, in contrast to most species of boa, which are typically viviparous. The females of many pythons are able to incubate their eggs using muscular contractions of the body to generate heat – an unexpected ability for a group of animals that are generally considered ectothermic and incapable of producing notable metabolic heat (see p.27). This is achieved by spasmodic contractions of the muscles, which has the effect of increasing the temperature between the female's body and the eggs around which she is coiled. When the ambient air temperature is too low, the female envelops the clutch more tightly within her coils and increases the rate of contractions, or 'twitching', while, if it is too high, she relaxes her body to permit greater ventilation, and twitches less frequently. A female python may stay coiled around her eggs for the entire incubation period (up to three months or more in some species), leaving them only occasionally to drink or bask.

ABOVE Children's python, *Antaresia childreni*, of Australia. The females of many python species brood their eggs for up to 3 months or more, during which time they do not feed and may leave only occasionally to bask or drink.

African and Asian pythons

Of the four species of python found in Africa, the most widely distributed and by far the largest is the African rock python, *Python sebae*. These huge snakes occur in dry bush country and forests, and, though mostly terrestrial, they climb well and are also semi-aquatic. Among their favourite haunts are riverbanks, from where they can slip away into the water when danger threatens. The royal python, *P. regius*, of West Africa, and the Angola python, *P. anchietae*, of the northern half of southwest Africa, are smaller, terrestrial snakes that seldom exceed 2 m (6½ ft) in length. Named for its handsome and clear-cut markings, the royal python is also called the ball python in allusion to its habit of rolling into a tight ball when threatened, ensuring that its head is protected in the centre. Royal pythons occur

ABOVE The African rock python, *Python sebae*, is exceeded in size perhaps only by the reticulated python. Its diet includes animals up to the size of small antelopes, and there are authentic reports of it also having eaten humans.

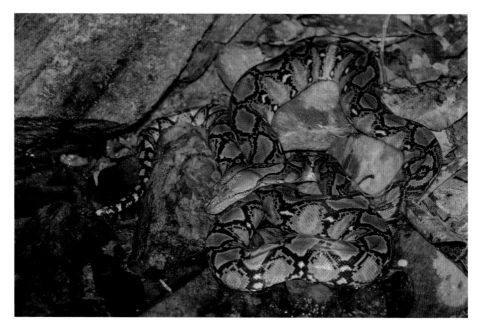

ABOVE Reticulated python, *Malayopython reticulatus*. Although primarily terrestrial, this huge Southeast Asian python is also an excellent swimmer and was one of the first vertebrates to re-colonize the volcanic island of Krakatau after its destructive eruption in 1883.

mostly in open forests and grasslands, and in parts of its range the species is quite common, whereas the Angola python is a comparatively rare species apparently more or less confined to rocky habitats. The fourth species, the Southern African python, *P. natalensis*, is distributed across much of southern Africa.

The longest species of snake in the world is the reticulated python, *Malayopython reticulatus* (see p.89), found in much of equatorial Southeast Asia, including the Philippines and parts of the Indonesian archipelago. Some individuals may attain lengths of 10 m (33 ft). It occurs in lowland areas, and, although essentially a terrestrial species, it climbs readily and is also often found in water. Another large, mostly terrestrial species is the Asian rock python, *P. molurus*, found over much of the Indian subcontinent and areas of Southeast Asia. Like the reticulated python it is an opportunistic ambush predator that may also actively forage for prey. Although much smaller, short-tailed or blood pythons are distinctive in having a particularly stout, heavy-set body. Previously regarded as a single species with geographically distinct populations (subspecies), these strikingly marked snakes are more usually now considered four separate

ABOVE Sumatran short-tailed python, *Python curtus*. Short-tailed pythons are terrestrial but also spend long periods immersed in muddy swamps or concealed among aquatic vegetation, where they lie in ambush mostly for small mammals.

species, all similar in body proportions and general habits, but varying in scale features, colour patterns and genetics. Adult Sumatran short-tailed pythons, *P. curtus*, are often dark, whereas the general body colour of the Borneo species, *P. breitensteini*, is more usually a rich yellow-brown. The southern mainland species, *P. brongersmai*, is sometimes bright red. The Burmese species, *P. kyaiktiyo*, has more ventral scales, occurs much further north than any of the other species, and was formally described only in 2011.

Australian and New Guinea pythons
Many different pythons are widely distributed in Australia, New Guinea and the many archipelagos in this region. Among those from Australia, four species of *Antaresia* are small snakes that rarely exceed 1.5 m (5 ft), and at less than 50 cm (20 in) long, *A. perthensis* from the western Pilbara region is the smallest of all pythons. Two other Australian pythons, comprising the genus *Aspidites*, differ from all others in lacking externally visible heat-detecting pits. The woma, *A. ramsayi*, occurs in arid habitats throughout much of the continent, while

ABOVE Macklot's python, *Liasis mackloti*, a large semi-aquatic species from Indonesia, Papua New Guinea, and coastal northern Australia.

the black-headed python, *A. melanocephalus*, is known mostly from northern Australia. These snakes mainly eat reptiles, including other snakes.

Largest of the pythons of the Australo–Papuan region is Kinghorn's python, *Simalia kinghorni*, found in northeast Australia and nearby islands, growing to lengths of 6 m (20 ft) or so. Another Australian 'giant', found only on the dry sandstone escarpments of Arnhem Land in northern Territory, is the Oenpelli python, *Nyctophilopython oenpelliensis*, famed for having remained undiscovered by scientists until the late 1970s. Adults may attain lengths of over 5 m (16 ft), and can overpower prey up to the size of a small wallaby. The Australian olive python, *Liasis olivaceous*, reaches a similar size and regularly eats wallabies. From the lowland monsoon forests and flooded savannas of New Guinea, the Papuan python, *Apodora papuana*, is a similarly large species, while Boelen's python, *Simalia boeleni*, also found only in New Guinea, is one of the most rarely seen of all pythons. This spectacular-looking snake is black with an overlying purple-blue sheen and a series of bright yellow diagonal streaks. It occurs in highland forests above 1,000 m (3,300 ft), but little is known of its habits in the wild.

Among various other pythons from the Australo–Papuan region, D'Albertis' python, *Leiopython albertisii*, is a terrestrial species of lowland rainforests in New Guinea. The green tree pythons, *Morelia viridis* and *M. azurea*, are exclusively arboreal, green snakes remarkably similar in appearance to the emerald tree boa in South America (see p.85).

THE REMARKABLE PHYSIOLOGY OF PYTHONS

It is well known that several pythons are able to consume enormous prey relative to their body size, a useful adaptation for animals that feed rarely. Because these snakes generally wait for their prey to come to them, they may have to go for long periods of time (up to several months) without feeding. Between meals, energy is scarce and pythons have evolved an amazing physiology that is able to save valuable energy during periods of fasting yet maximize the use of food resources when they appear or are needed.

A recent study of the Burmese python, *Python bivittatus*, has shown that while waiting for its next prey item, the digestive and blood circulatory systems are vastly reduced in size, structure and function. The stomach, pancreas and gallbladder do not produce or secrete their usual chemicals, enzyme activity is depressed and the intestine becomes greatly reduced. Even the heart, liver and kidneys have a much-reduced mass. These adaptations are thought to have helped the invasive population of Burmese pythons in Florida survive freezing conditions.

What is perhaps most startling, however, is the speed with which the dormant viscera are able to resume their normal function after prey has been caught. In only 24 hours, acid production by the stomach reduces the pH from neutral (pH 7) to very acidic (pH 2). This dissolves the meal, so that by day three all that is left of a mammal is some vertebrae and clumps of hair. In preparation for this broken-down matter (chyme) from the stomach, the microvilli (tiny finger-like projections on the walls) of the small intestine (which take up nutrients into the bloodstream) grow at an impressive rate, doubling in length in the first six hours. The mass of the small intestine increases by 70% within 24 hours of feeding and within 48 hours metabolic rate can increase 40-fold. After six or seven days the system begins to down-regulate once more, also with impressive speed.

In total the python's liver and kidneys can double in size and breathing and heartbeat output increase five-fold while digesting. Even this large increase in breathing is not sufficient to meet the increase in gas exchange required to fuel digestion and so the snake hypoventilates during this period.

The benefit of largely shutting down the gastrointestinal system when no food is available is having a much-reduced metabolic rate, saving precious energy. Snake species that shut down their digestive systems in this way show an almost 50% reduction in metabolic rate when compared with species that feed more regularly and maintain a similar level of activity in their guts when fasting.

FAMILY BOIDAE: True Boas

ABOVE Common boas, *Boa constrictor*, are among the largest snakes in tropical America, although even large specimens, such as this 3 m (10 ft) long example at the side of a forest track, are not easy to see.

SUBFAMILY BOINAE: Constrictors, anacondas, and tree, rainbow and Caribbean boas

Perhaps the most familiar of all the snakes in this group is the common boa constrictor, *Boa constrictor*. This tropical American species has one of the largest geographic distributions of any boid, ranging throughout much of South America and the Caribbean island of Trinidad and Tobago, occurring in a wide range of habitats. Recently thought to be the only species in this genus, *B. constrictor* has now been split into five species, the other four having much more restricted ranges in Central America (*B. imperator, B. sigma*) or some Caribbean islands (*B. nebulosa, B. orophias*). These boas are large and heavy-bodied, and although they were long believed to have reached a length of 5.6 m (18½ ft), they rarely seem to grow larger than 3.6 m (12 ft).

Another famous group of boas are the anacondas (*Eunectes*) of which the green anaconda, *E. murinus*, is arguably the largest (but not longest) snake in the world (see p.89). This truly enormous species has a wide distribution over much of tropical South America. Three smaller species of anaconda also occur in

ABOVE A new-born emerald tree boa, *Corallus caninus*. As these snakes mature their colour pattern gradually changes to bright green (see p.85).

South America, the yellow anaconda, *E. notaeus*, occurs in southern parts of the Amazon basin, the dark-spotted anaconda, *E. deschauenseei*, lives on the island of Marajó in the mouth of the Amazon, and the Beni anaconda, *E. beniensis*, is from Bolivia. Anacondas are highly aquatic boas in which the eyes and nostrils are on top of the head, and they are almost always found in or near to water, especially swamps and slow-moving rivers. They are mostly ambush-hunters and feed on a wide range of prey. In the seasonally flooded llanos (grassy plains) of Venezuela, green anacondas conceal themselves beneath dense floating mats of water hyacinth, from where they catch prey as large as the sheep-sized capybaras and occasionally small caimans.

Nine species of American tree boa in the genus *Corallus*, are a specialized group of tree-dwellers, most of which have relatively slender bodies, long prehensile tails, and very long mandibular teeth. The genus is distributed throughout Central and South America and several Caribbean islands. Cropani's boa, *C. cropanii*, may be one of the rarest snakes in the world (see p.74). Tree boas are nocturnal snakes that use both active and ambush hunting methods to catch their prey. In terms of food preferences, the least specialized is the common tree boa, *C. hortulanus*, which eats frogs, lizards, birds and small mammals, including bats. Perhaps the most recognizable of *Corallus* species is the emerald tree boa, *C. caninus*, due to its bright green appearance, and their colour-change abilities from juveniles to adults, with juveniles varying from bright red to yellow. Emerald tree boas and annulated tree boas, *C. annulatus*, are relatively stout-bodied snakes that show a strong preference for rodents and other endothermic prey. Grenada tree boas, *C. grenadensis*, feed on lizards and small mammals. Some of the densest populations of this species are found in or next to fruit orchards, due to high prey abundance.

Five species of rainbow boa can be found throughout much of Latin America. The most widely distributed species is *Epicrates cenchria*, which occurs throughout the biogeographic region of Amazonas. The Argentine rainbow boa, *E. alvarezi*, is the most genetically distinct of this group and is found in the Chaco woodlands of Argentinia, Paraguay and Bolivia, while the brown rainbow boa, *E. maurus)* inhabits dry forests from Nicaragua to northern Brazil. *Epicrates assisi* primarily occurs in xerophytic forests of the Caatinga and other areas of northeastern Brazil. *Epicrates crassus* is distributed from the Andean slopes of Bolivia to the east coast of Brazil and south to Argentina.

ABOVE Rainbow boa, *Epicrates cenchria*, from tropical South America, Paraguay and Argentina.

ABOVE The Cuban boa, *Chilabothrus angulifer*, is an inhabitant mostly of forested areas and, although typically found in trees, is equally at home on the ground.

The Caribbean boas of the genus *Chilabothrus* were, until recently, classified in the same genus as the rainbow boas (*Epicrates*). At a length of more than 3 m (10 ft), *C. angulifer* from Cuba is by far the largest of the 14 species in the genus. These snakes are semi-arboreal in habit and feed on a wide variety of prey. Cuban boas (*C. angulifer*) and Puerto Rican boas (*C. inornatus*), in particular, are known for their habit of catching bats as they emerge at dusk from cave entrances. Among the smaller species, Haitian vine boas (*C. gracilis*) occur in lowland woods near water, whereas the Mona Island boa (*C. monensis*) is found in dry habitats. Pregnant females of the Mona Island boa have been found in termite nests and may use these sun-baked places for regulating their body temperature.

SUBFAMILIES SANZINIINAE AND CALABARIINAE: Madagascan and African boas
There are four species of Boa found on Madagascar. The Madagascan ground boa (*Acrantophis madagascariensis*) and Dumeril's ground boa (*A. dumerili)* are stout, terrestrial species that attain lengths of approximately 2 m (6½ ft), while the Madagascan tree boa (*Sanzinia madagascarensis*) and the Nosy Komba tree boa (*S. volontany*) are slightly smaller and more slender tree-dwellers.

The closest living relative of the Madagascan boas, the Calabar boa, *Calabaria reinhardtii*, inhabits forested regions of West Africa. Because of its habit of laying eggs, this unusual snake was long thought to be related to pythons and only recently have biologists found evidence (particularly from DNA) suggesting it has closer affinities with boas. When threatened with danger, Calabar boas tend to raise the tail to divert attention away from the head, keeping the latter tucked safely beneath their body coils. They may also roll themselves into a tight ball and release a potent musky scent from special anal glands. These are cylindrical, burrowing snakes up to approximately 1 m (39 in) that have exceptionally thick and toughened skin, possibly to protect them from bites when they raid nests of rodents to feed on their young.

ABOVE Dumeril's ground boa, *Acrantophis dumerili*, is one of four boa species restricted to Madagascar.

'There lay in the mud and water, covered with flies, butterflies and insects of all sorts, the most colossal anaconda which ever my wildest dreams had conjured up. Ten or twelve feet of it lay stretched out on the bank in the mud. The rest of it lay in the clear, shallow water, one huge loop of it under our canoe, its body as thick as a man's waist. It measured fifty feet for certainty, and probably nearer sixty'.

So wrote an explorer (F W Up de Graff, in *Head Hunters of the Amazon*, 1923) of an encounter in Amazonian Ecuador with a huge anaconda – probably greatly exaggerated because there are no reliable records of snakes greater than 8.7 m (28½ ft) in length. Although possibly not the longest snake in the world, the green anaconda, *Eunectes murinus*, is certainly the most massive. Most early references to 'giant' anacondas and pythons appear to have been grossly exaggerated or based on the length of the removed skin, which is easily stretched. For example, the longest known anaconda is often reported to be 11.4 m (37½ ft), found in eastern Colombia during the 1930s, but the accuracy of this measurement has been doubted and truly reliable records indicate that the maximum size attained by these snakes is closer to 8 m (26 ft).

Another 'giant' is the reticulated python, *Malayopython reticulatus*, of Southeast Asia. This species is also claimed to attain lengths of 9 m (30 ft) or more, though it does not rival the girth and bulk of its South American relative. The most reliable documented length for a reticulated python is for an individual appropriately named 'Colossus', that lived for many years in USA's Pittsburgh Zoo; it measured 8.7 m (28½ ft) and weighed 145 kg (320 lb). Almost as controversial as the size attained by anacondas and the larger species of python are tales about what they reputedly eat. In parts of South America, for example, anacondas are

ABOVE Green anaconda, *Eunectes murinus*, at Ilha Caviana on the Amazon River, Brazil.

often held responsible for the unexplained disappearance of horses and oxen, and in Africa large rock pythons, *Python sebae*, are believed by some to prey on buffalo. In 1952 a Sri Lankan newspaper even carried an article describing a python (presumably *Python molurus*) attacking a baby elephant. Such stories seem outrageous, and are probably based on little more than hearsay and fabrication, but these snakes are nonetheless capable of consuming enormous meals. One genuine case concerns a 5.5 m (18 ft) Asian rock python that, having been discovered with a huge bulge in its stomach, was found to have eaten a full-grown leopard. They need a remarkable physiology to process such huge meals (see p.83).

Although today's anacondas and pythons can grow to impressive sizes, they would be dwarfed by some of the extinct snakes known only from fossil remains. The largest snake ever known to have lived is the anaconda-like *Titanoboa cerrejonensis* (see p.14), fossils of which were discovered in Colombia from rocks approximately 60 million years old. This colossal snake is estimated to have reached a length of at least 13 m (43 ft) and a mass of more than 1,000 kg (2,200 lb). The climate must have been notably warmer at that time for snakes to have reached this size.

SUBFAMILY CANDOIINAE: Pacific boas

Five species of Pacific boas in the genus *Candoia* are unusual in that they occur in the area of New Guinea and Pacific islands to the north and east, almost as far away from other boas as is possible. Their dorsal body scales are strongly keeled, rather than only partially keeled or smooth as they are in most other boas, and they differ also in having a flat, angled rostral scale that gives the snout a distinctly oblique profile. The New Guinea ground boa, *Candoia aspera,* is a small, stout, semi-burrowing inhabitant of the forest floor, whereas *C. carinata* is larger and slimmer and, although mostly terrestrial, is also found in trees. *Candoia bibroni,* the largest and most slender of the Pacific boas is almost exclusively arboreal and has a surprising distribution for a single species, occurring across Pacific islands that are geographically far apart.

ABOVE The ground boa, *Candoia paulsoni mcdowelli,* from New Guinea.

SUBFAMILY CHARININAE: Rosy and rubber boas

The genera *Charina* (rubber boas) and *Lichanura* (rosy boas) are closely related, small genera, both containing only two species, occurring in North America. They share burrowing habits and are relatively short, rarely growing over 1 m (3½ ft) long. The northern rubber boa, *Charina bottae,* is found further north than

ABOVE The rubber boa, *Charina bottae*, an example from California, USA.

any other boa or python, even ranging into southwestern Canada. This species appears to be especially tolerant of cold. Active specimens have been measured with body temperatures of less than 7°C (44°F), although these snakes are unable to withstand the freezing temperatures of winter months and so hibernate underground during that period. The rubber boas occur in many habitats, from dry grasslands to humid woodlands and mountain forests, and may be found at elevations above 3,000 m (9,800 ft). A second species, the southern rubber boa, *C. umbratica*, was recently considered a distinct species from *C. bottae*; its range is restricted to California. The rosy boas occur in drier, desert areas of California and northwestern Mexico.

Rosy boas occasionally forage for food above ground, and rubber boas have even been known to climb trees to raid birds' nests, but they generally hunt and feed in underground tunnels. The feeding habits of the rubber boa are unusual but are shared by the West African Calabar burrowing boa (see p.88). They capture and asphyxiate one or two small mice by trapping them with a coil of the body against the walls of their burrows while simultaneously swallowing another. At the same time they use their short, stubby tail, the terminal vertebrae of which are fused into a bony club, to fend off attacks from the mother mouse. To divert her attention, the snake will even allow her to chew on its tail, so it is perhaps not surprising that the tails of rubber boas found in the wild are almost always heavily scarred.

SUBFAMILY UNGALIOPHIINAE: Banana and Oaxacan boas

Two genera of small, nocturnal boas live on the forest floors of Central America. *Ungaliophis* and *Exiliboa* were once thought to be closely related to the tropidophiids (dwarf 'boas', see p.65), and they do share several features of anatomy and behaviour, but are now known to belong in the true boa family while dwarf 'boas' are a distinct lineage. *Ungaliophis* contains two species with separate distributions in Central America, and the single species of *Exiliboa* (*E. placata*) is restricted to high-elevation cloud forests of Oaxaca, Mexico. The configuration of the head scales in these species is particularly distinctive. In *Ungaliophis*, the prefrontal scales are coalesced and expanded over much of the snout, whereas in *Exiliboa*, which has the more usual condition of paired prefrontals, it is the internasals that are fused, forming a single, large, triangular-shaped plate.

Occasional specimens of *Ungaliophis* reaching the USA in shipments of bananas led to them becoming commonly known as 'banana boas'. Their maxillary teeth are somewhat specialized and are perhaps adapted for feeding on small, tree-dwelling frogs and lizards, their main prey. The northern species, *U. continentalis*, lives mainly at low to intermediate elevations from Chiapas in southern Mexico, to Honduras, and can be found above 2,000 m (6,600 ft)

ABOVE Banana boa, *Ungaliophis panamensis*.

in the pine forests of eastern Chiapas. *Ungaliophis panamensis*, the southern species, is found from southeastern Nicaragua through Costa Rica and Panama into northern Colombia. Adult *Ungaliophis* attain an overall length of about 70 cm (28 in). They are nocturnal and mainly arboreal.

The Oaxacan boa, *Exiliboa placata*, is found only in the cool, moist forests that prevail at 2,000–3,000 m (6,600–9,800 ft) on the mountain slopes of Oaxaca, Mexico. The generic name *Exiliboa* comes from the Latin 'exigere', meaning 'to banish', in allusion to its isolated occurrence. A small snake, perhaps reaching no more than 50 cm (20 in), its body is compressed towards the base of the tail, where it is appreciably higher than wide, and external vestiges of limbs are present in both sexes. The dorsal surface is almost uniformly black. Specialized for burrowing in wet leaf litter, it preys mainly on small frogs and salamanders and their eggs.

SUBFAMILY ERYCINAE: Sand boas

There are 13 species of sand boa, which are sometimes grouped all together within the genus *Eryx*. They are small burrowers that occur over much of Asia and parts of northern Africa, they are particularly suited to life in sandy habitats by having the following adaptations:

ABOVE The European javelin, or spotted sand boa, *Eryx jaculus*.

ABOVE Rough-scaled sand boa, the South Asian *Eryx conicus*.

- The skull is compact, and the small conical head has a rounded snout for burrowing into the sand head-first. In some species the lower jaw is also strongly countersunk.
- The body is cylindrical with a short tail.
- Many species have a hardened, horizontal ridge across the snout.
- The nostrils are on top of the snout, and in many species the openings have been reduced to narrow slits preventing earth and sand being drawn into the respiratory tract.
- Physiological adaptations enable them to reduce water loss and withstand the hot, dry conditions under which many live.

Like the rosy and rubber boas, most sand boas hunt for prey within subterranean tunnels, although some may also ambush prey by hiding in loose sand just beneath the surface, striking upwards at small animals that stumble over them. Most species eat small rodents and lizards, seizing them with a rapid, sidelong, slashing bite.

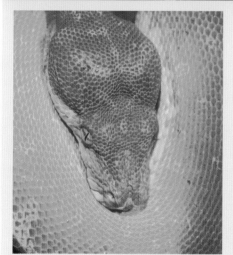

ABOVE Green tree python, *Morelia viridis*.

LEFT Emerald tree boa, *Corallus caninus*.

Some snakes living in different parts of the world and that are not especially closely related nonetheless bear a striking resemblance to each other. A clear example of this phenomenon is that of the green tree python, *Morelia viridis*, found in Australia and New Guinea, and the emerald tree boa, *Corallus caninus*, of South America. These species are both tree-dwellers that change from yellow or brown as juveniles to bright green as adults. They are the same general size and shape, have similarly shaped heads with greatly enlarged anterior teeth, and also share the same style of resting in trees, looping their bodies over a horizontal branch. The reason why they are so similar, however, is not because they are especially closely related (they are classified in two different families), but because they have similar lifestyles and have both adapted to live in similar environments. Over the course of evolution they have been 'moulded' into similar-looking species by the same selective pressures.

Other examples of strongly convergent evolution of ecology and morphology in snakes include the similarly low and wide head shapes evolved independently in natricine species that forage in sediments in aquatic habitats, and multiple instances of small-headed ("microcephalic"), narrow-bodied sea snakes whose disparate morphologies have evolved as adaptations to feeding on burrowing eels.

ACROCHORDIDAE: Asian and Australasian File Snakes

The Asian and Australasian file snakes, or 'wart snakes', of the family Acrochordidae are so called because of the coarse, granular appearance of their scales. Acrochordids are not to be confused with African 'file snakes', (see p.155), that also have rough scales but not of a granular nature. Acrochordids are one of the most distinctive families of snakes, but comprise only one living genus (*Acrochordus*) with three species. In size and body proportions these unusual snakes resemble some pythons and boas, although in other features (such as a single lung, absence of a pelvis and hindlimbs, and aspects of the skull and jaw bones) they are more similar to many other caenophidian snakes. Recent analyses of DNA have confirmed results from morphological studies that indicate that acrochordids are the sister group to all other caenophidian snakes, i.e., they form one half of the oldest divergence in the caenophidian evolutionary tree (see p.52).

Recent studies have indicated that living acrochordids evolved over approximately the last 20 million years. The three living species occur from coastal India through the Thai-Malay peninsula and into New Guinea and northern Australia. Fossil acrochordids are known from deposits as old as 18 million years ago and are especially well-known from localities in Pakistan and northern India. This is further inland and north and west than the distribution of the living species, but the extinct species resemble them nonetheless in that they also lived in aquatic environments.

Acrochordids are entirely aquatic. They occur mostly in estuarine and freshwater habitats, although the smallest acrochordid, *Acrochordus granulatus*, no more than about 1 m long) may also be found in coastal marine waters as far as 10 km (6¼ miles) offshore, often in areas frequented by true sea snakes. Much like sea snakes (see p.143), acrochordids have acquired a series of external and internal modifications that enable them to take full advantage of their aquatic surroundings. The only other major lineage of highly aquatic and at least partially marine snakes are the Asian mud snakes Homalopsidae (see p.119).

The most conspicuous feature of acrochordids is their skin, which is loose and baggy, and covered with small, tubercular (wart-like), non-overlapping scales. As a result, common names for *Acrochordus* in Malaysia and Indonesia include 'ular karung', 'ular kadut' and 'ular kain', which all mean sack or cloth snake. The largest species, the Javan file snake, *A. javanicus,* can exceed 2 m (6½ ft) in length and is also sometimes called 'elephant's trunk snake' in these countries. There are

ABOVE The largest of the three living species of *Acrochordus*, the Javan file snake, *A. javanicus*.

no broad, transverse scales on the abdomen, and the scales on the head are also all of the same general shape and form. Few other large snakes have such small, evenly sized scales on both the dorsal and ventral surfaces of the body, and this particular quality has made the skin of *Acrochordus* prized in the leather trade. In Southeast Asia, many tens of thousands of specimens are probably harvested directly from the wild for this purpose every year (see p.123).

Acrochordids feed exclusively on fish. They may seize prey with a sudden sideways snap and subdue larger fish with the assistance of body loops. The coarse, 'gritty' nature of the scales may help restrain slippery, struggling prey. Observations of wild, free-ranging Arafura file snakes, *A. arafurae*, show that these snakes have very low feeding rates and may eat prey only a few times each year. Acrochordids are viviparous. Female *A. arafurae* produce litters of up to about 25 offspring, although this species reproduces less frequently than do most other snakes, perhaps as little as once every 10 years.

ABOVE The Arafura file snake, *Acrochordus arafurae*, an example from New Guinea. Note the granular scales, dorsally positioned eyes, and nostrils at the tip of the snout.

Acrochordids have an unusually low rate of metabolism (only about half that of other snakes) and appear incapable of sustained activity for more than a few minutes. Their low metabolic rates are also associated with low reproductive rates. There are differences among the three species, but *A. arafurae* females might reproduce only once every few years. Out of the water they are generally sluggish and seem almost helpless. Studies of *A. arafurae*, however, have shown that these snakes, although slow in their movements, are capable of travelling for considerable distances in the seasonally flooded billabongs of northern Australia where they occur. During the day they remain hidden under overhanging banks, beneath sunken logs, or amongst waterweed, but while searching for food at night they frequently cover distances of several hundred metres.

In addition to their remarkably low metabolism, acrochordids are highly unusual in other regards. They have tiny hair-like projections on their scales, particularly dense on the head, that sense movement in the often murky water they live in. The little file snake, *A. granulatus,* have very high volumes of blood for their size, and their blood also carries a higher concentration of red blood cells than in other snakes. These features combine to allow them to hold a large amount of oxygen, preventing the need to come to the water surface frequently to breathe.

XENODERMIDAE: Strange-skinned Snakes

The family Xenodermidae comprises a poorly understood, fairly small lineage of snakes. They are relatively little-studied, and historically there was great uncertainty as to which snakes should actually be classified within this family. The family currently includes six genera and 30 species from Southeast Asia. More than half of these species have been described as new to science this century. In general, xenodermids are rather peculiar-looking snakes with long bodies, distinctively enlarged heads, long tails, and, in some species, oddly formed scales. Among the more widely distributed and best known is the dragon snake, *Xenodermus javanicus*, from Thailand, peninsular Malaysia, Java, Borneo and Sumatra. This unusual little snake, only about 70 cm (28 in) long, has granular dorsal scales, with three rows of large, keeled tubercles. On its lower sides the scales are triangular and separated by areas of bare skin – which gives the snake its name, with 'xeno' meaning strange and 'dermis' skin. A frog-eater, it lives in wet leaf litter or waterlogged soil in tropical forests, swamps, marshes and rice paddies, where for much of the time it leads a semi-aquatic or partly burrowing existence. *Xenodermus* has most often been encountered by biologists who conduct fieldwork at night, when it can be found actively foraging on the surface. The other members of the Xenodermidae occur in Borneo and into Indochina, Japan and northeast India. The majority, 22 species, are classified in the genus *Achalinus*.

ABOVE An example of the xenodermid *Xenodermus javanicus*, found at night on the edge of a small rocky stream in a rainforest in Java.

PAREIDAE: Asian Slug Snakes and Wood Snakes

There are currently 41 species of snake recognized in the family Pareidae, classified in the genera *Aplopeltura*, *Asthenodipsas*, *Pareas* and *Xylophis*. The 36 species classified in the former three genera comprise the subfamily Pareinae, found throughout East and South Asia. Pareine pareids subsist largely on terrestrial gastropod molluscs, giving rise to the common name snail or slug snakes. The dietary specialization of pareids is shared with several neotropical snakes in the Dipsadinae subfamily of Colubridae (species of *Dipsas*, *Sibon* and *Sibynomorphus*) and, despite their different evolutionary origins, they resemble each other closely in appearance (see p.177). Most snail-eating snakes are small – less than 1 m (3¼ ft) in length. The head is often large, with conspicuously large eyes, and the snout is usually short and blunt. They are adapted for climbing, with long bodies flattened from side to side, and the head well differentiated from the slender neck. Pareines eat only the soft body of the snail inside the shell; grasping the body

BELOW The pareine pareid *Aplopeltura boa*, photographed in Malaysian Borneo.

with its needle-like teeth, the snake extends its jaws alternately from side to side and continues advancing its grip in this way until the mollusc is dragged bodily from its shell. Interestingly, some snail-eating pareines have been found to have asymmetrical jaws, with more teeth on the right lower jaw than on the left. This is thought to aid them while foraging, because the vast majority of snails display dextral (clockwise) whorls. In experiments where the Japanese *Pareas iwasakii* were fed artificially bred sinistral (anti-clockwise) and dextral prey, they were able to extract the body of dextral snails from their shells faster than sinistral ones, and dropped the sinistral snails more often. Some snail-eating aquatic arthropods also have asymmetrical feeding apparatus. Two species of snail-eaters from Africa (of the pseudoxyrhophiid genus *Duberria*) occasionally deal with larger snails by smashing their shells on the ground, as some birds do, but these species are not closely related to the Asian pareid or New World dipsadine snail and slug eaters.

The remaining six species of pareids comprise the subfamily Xylophiinae and genus *Xylophis*. These are all small, burrowing snakes restricted to high-rainfall areas of southwestern peninsular India. With their small eyes, generally cylindrical bodies, short tails and narrow heads (see p.54), *Xylophis* do not superficially resemble the other (pareine) pareids, and only in 2018 was DNA evidence discovered that strongly indicated they should be classified in the same family.

VIPERIDAE: Vipers

Vipers comprise the family Viperidae, which includes nearly 400 species of highly specialized snakes. They have a characteristic venom-injecting apparatus that is more sophisticated than that found in any other group. Unlike mambas, coral snakes and other elapids, vipers have large, erectile fangs that are capable of being 'pivoted' independently of one another (see p.21). Normally, these are kept folded back along the roof of the mouth, encased in a protective sheath of soft tissue (the vagina dentis), but are rotated forwards as the mouth is opened to strike. The fangs have enclosed venom canals and their large size enables vipers to inject venom deep into the tissues of their prey, where it is rapidly absorbed; in this respect the fangs can be compared with a hypodermic needle. A West African species, the Gaboon viper, *Bitis gabonica*, is credited with having the longest fangs of all snakes, measuring around 5 cm (2 in). Aside from the vipers, only the stiletto snakes (atractaspidids of the genus *Atractaspis*, see pp.150–151) have similarly large, front-mounted, hollow and movable fangs, but in that group the articulation between the maxilla and the rest of the skull is uniquely developed into a ball-and-socket joint.

Most vipers are heavy-bodied, terrestrial snakes, and adaptation to different niches in the family has perhaps not been as extensive as in some others. With the exception of a few desert-dwelling species that are able to 'sink' vertically into loose sand by shuffling their bodies, no vipers are habitual burrowers. Only two North American species of cottonmouth, *Agkistrodon piscivorus* and *A. conanti*, are semi-aquatic. There are, however, many tree-dwelling species that have relatively slender bodies with strongly prehensile tails.

The Viperidae has an almost worldwide distribution and embraces three subfamilies: the night adders and 'true' vipers (Viperinae); the Azemiopinae, with only two species; and the pit vipers (Crotalinae). A Eurasian species, the common viper or adder (the viperine *Vipera berus*), has a particularly large range, including areas within the Arctic Circle, and occurs further north than any other snake. The snake with a range that extends further south than any other species is also a viper, the Patagonian lancehead, the crotaline *Bothrops ammodytoides*.

FEEDING HABITS OF VIPERS

Vipers are among the greatest users of the 'sit-and-wait' hunting technique, and many have evolved highly cryptic body markings that enable them to remain concealed from both their predators and prey. Using chemical traces of their prey to seek out strategic ambush sites, they lie in wait for a meal to pass within striking range. They remain motionless in the same place, often for days or even weeks, until either their patience is finally rewarded or instinct drives them to try elsewhere.

In some vipers the tip of the tail is coloured and/or shaped differently and used as a lure for attracting prey. Peringuey's adders, the viperine *Bitis peringueyi*, buried in desert sand, will wriggle their black-and-white-banded tail tips above the surface to attract foraging lizards, and the spider-tailed horned viper, *Pseudocerastes urarachnoides*, (see pp.109–110) is even more remarkable, yielding an exceptional, spider-shaped tail. Similar luring behaviour is seen in some non-viperid snakes, such as the death adders (*Acanthophis*, Family Elapidae) from Australia and New Guinea.

The bite of a viper takes the form of a rapid 'stabbing' movement, for which the fangs are fully erected and the mouth opened to almost 180° (see p.21). If the prey is small, some vipers may restrain it in their mouth until the venom

ABOVE Mexican cantil, *Agkistrodon bilineatus*. Juveniles of this Central American crotaline viperid have bright yellow tail-tips and flick them about in the manner of a wriggling worm to entice prey within striking range.

takes effect, but in many cases the fangs are withdrawn immediately, and the dying animal is located afterwards by following its scent. Throughout a snake's life the fangs are continuously replaced by others that grow and move forward from behind; when the next-in-line is ready for use, the functioning fang loosens at its base and either falls out or is left embedded in the body of the snake's next meal (and then often consumed by the snake). The venom of many vipers seems to have a digestive as well as immobilizing effect on their prey. This has been suggested to be a major contributory factor in the ability of some viper species to live in seasonally cold environments, where low temperatures may otherwise restrict effective digestion.

SNAKEBITE AS A NEGLECTED TROPICAL DISEASE

Snakebite envenoming has recently been identified as a Neglected Tropical Disease (NTD) by the World Health Organization (WHO). Not only does snake envenoming cause the deaths of 81,000–138,000 people each year but it leaves thousands more with life-changing injuries and crushing debt. The majority of serious snake envenoming occurs in rural areas of some of the poorest countries in the world, where medical treatment is sparse or expensive. The species that cause the highest number and most serious snakebites in the world are Russell's viper (Asia), cobras (Africa and Asia), saw-scaled vipers (Africa and Asia), puff adders (Africa), Malayan pit vipers (Asia), mambas (Africa), rattlesnakes (Americas), lance-headed vipers (Americas), brown snakes (Australasia), taipans (Australasia), and kraits (Asia).

That the WHO has recognized snakebite as a NTD provides new hope for many communities around the globe. The WHO has an ambitious plan to reduce mortality and disability by 50% before 2030. The best way to halt mortality and disability is through community education by ensuring proper safety precautions are adhered to, understanding how to treat snakebite (training in first aid to members of the general public, and in advanced treatment to medical professionals) and encouraging people to seek medical attention as soon as they are bitten. Many antivenoms (used to treat snakebite) are expensive and/or are not safe, and the WHO is working to try to create more cost-effective and better-quality treatments, improved healthcare capacity and enhanced coordination between organisations and governments.

ABOVE Scientists at the Evolutionary Venomics Laboratory, Indian Institute of Science, Bengaluru, India, studying the diversity and function of venoms to improve snakebite therapy.

ABOVE A snakebite awareness and education community event in Kitui County, Kenya.

SUBFAMILY VIPERINAE: Night Adders And 'True' Vipers
Night adders

The seven species of night adder, all within the single genus *Causus*, are found only in Africa, south of the Sahara Desert. Although night adders generally live up to their name, some species have been seen venturing out in broad daylight or basking in the morning and evening. The most widespread species is the rhombic night adder, *C. rhombeatus*, which occurs throughout the larger part of the continent. They have smooth scales, symmetrically arranged on the crown of the head, and eyes with circular pupils, and thus differ from most other vipers in overall appearance.

Unlike most other vipers, night adders lay eggs rather than giving birth. Typically less than 80 cm (32 in) long, most are brown or grey with patterns of spots or blotches but some species are more colourful, such as the velvety night adder, *C. resimus*, which is a beautiful green colour.

The snouted night adder, *C. defilippi* of East Africa, has an upturned nose and perhaps uses it for rooting out its prey. Although comparatively inoffensive, when thoroughly provoked night adders draw the body into a defensive coil, inflate themselves with air, and emit a surprisingly loud, guttural hiss. Night adders have large venom glands and their venom is especially toxic to toads, their natural prey, but appears to be rather less dangerous to humans than that of many other vipers.

TOP Velvety night adder, the viperine viperid *Causus resimus*, from Africa.

LEFT Rhombic night adder, the viperine viperid *Causus rhombeatus*. Night adders are African terrestrial snakes that feed largely on amphibians, especially toads.

True vipers

The 'true' vipers, of which there are approximately 92 species divided among 12 genera, are restricted to the Old World regions of Europe, Asia and Africa. They differ most conspicuously from pit vipers (Crotalinae) in lacking facial heat-sensitive pits.

Twenty-one species of the genus *Vipera* occur mostly in Europe and parts of Asia, where in places (such as Britain) they represent the only venomous snakes. They are all fairly heavy-bodied snakes with short tails, well-defined, triangular heads, relatively large head scales, and distinctive zigzag patterns along the body. Some species, such as the nose-horned viper, *V. ammodytes*, have a scaly 'nose-horn' on the tip of the snout. Terrestrial and ground-dwelling, they occur in a variety of habitats including open forests, sandy heaths, wet meadows, and dry, rocky hillsides.

Two particularly widespread species, the European viper or adder, *V. berus*, and the asp viper, *V. aspis*, are found at elevations of up to about 3,000 m (9,800 ft) in the European Alps. All species are active by day, although at least some become partly nocturnal when night-time temperatures are high enough. Most hibernate during winter, often in communal dens (hibernacula), and after emerging in spring some migrate short distances to different feeding grounds. Although most feed on rodents and lizards, the small meadow viper, *V. ursinii*, eats mainly insects. Two similar species in the northern African and southern Asian genus *Macrovipera* and four in the widespread Asian genus *Daboia* are larger snakes characterized by small head scales.

RIGHT With a scaly horn on the tip of its snout, the nose-horned viper, *Vipera ammodytes*, from the eastern Mediterranean region is one of the more distinctive European and western Asian vipers.

The adder, the viperine viperid *Vipera berus*, is one of the world's most successful snakes. It has the largest geographical distribution of any terrestrial land-living species, ranging from the British Isles, across Europe and northern Asia, east to the Pacific Ocean, and also occurs further north than any other snake. Adders exhibit considerable colour dimorphism; top, year-old juvenile; centre left, reproductively active male; bottom left, female.

RIGHT Although not a pit viper, the facial nerve endings in Russell's viper, the Asian viperine *Daboia russelii*, are highly sensitive to temperature variation. Several other snakes that lack externally visible pits also have heat-sensitive areas on their heads, including puff adders (*Bitis arietans*) and *Boa constrictor*.

Among the most dangerous of all snakes to humans are the 12 species of saw-scaled or carpet vipers (*Echis*) from North Africa, the Middle East and Asia, which are responsible for tens of thousands of snakebite fatalities and morbidities each year. When alarmed, these hot-tempered little snakes rub the coils of their bodies together, producing a curious rasping sound that serves to warn off many potential predators without the snake having to resort to wasting venom (a metabolically expensive resource).

The 18 species of African bush vipers (*Atheris*) are distinctive-looking snakes with short, rounded heads and large eyes. Some are brightly coloured and have exceptionally fringed scales. Found in the equatorial forests of Africa, most are tree-dwellers. The more cryptically coloured *A. barbouri* from Tanzania's Udzungwa mountains is ground dwelling, found in grassland or among leaf litter, and feeds on earthworms. Two other African species, Hindi's viper (*Montatheris hindii*) from high-elevation moorlands in Kenya, and the lowland swamp viper, *Proatheris superciliaris*, from Mozambique, Malawi and Tanzania, are also ground-dwelling.

The 18 species in the exclusively African and Arabian genus *Bitis* are exceptionally stout-bodied vipers with broad, flattened, triangular heads. All are strictly ground-dwelling. With a distribution extending throughout Africa and into southern Arabia, the puff adder, *B. arietans*, is the most common and widespread, and is responsible for many serious snakebites. The massively built Gaboon viper, *B. gabonica*, is the largest, attaining lengths of almost 2 m (6½ ft) and weights of over 10 kg (22 lb). This formidable snake has been known to eat small antelopes and even porcupines. Gaboon vipers and a similar-looking

ABOVE Great Lakes bush viper, *Atheris nitschei*, an east African viperine often found in the peripheral forests around Lake Victoria and other large lakes. The milky coloured eyes are a symptom of the skin-shedding process.

species, the rhinoceros viper *B. nasicornis*, are tropical-forest-dwellers with curious scaly horns on the tip of the snout and camouflaging colour patterns that resemble dead leaves. Although large and with a distinctive colour pattern of fine yellow reticulations on a black background, the Bale Mountains adder, *B. harenna*, was described as a distinct species only in 2016. It likely has a very restricted distribution, being found only within a small region of the Ethiopian highlands. This is a very poorly known species that has thus far been studied only from a single museum specimen collected in the 1960s, and a brief opportunistic encounter with a live individual on a mountain road in 2013.

Other species, such as the horned adder, *B. caudalis*, are small desert-dwellers with conspicuous horn-like projections over the eyes, a feature they share with seven species of African and Asian sand vipers (*Cerastes* and *Pseudocerastes*). As discussed previously in this section, the spider-tailed viper, *P. urarachnoides*, found in Iran, has extremely elongated scales on the tip of its tail, which ends

ABOVE The Iranian spider-tailed viperine viperid, *Pseudocerastes urarachnoides*.

RIGHT Horned sand viper, *Cerastes cerastes*. This northern African and Arabian desert viperine buries itself in sand where it lies in wait for prey with only its eyes and nostrils visible.

in a bulbous growth. The feature bears a strong resemblance to a spider, or more precisely a 'sun spider' (order Solifugae – not true spiders), which live in the same habitat. The snake uses this ornament as a lure for bird prey and it even moves the tail in a jerky, arachnid-esque manner, which adds to the ruse.

Macmahon's viper, *Eristicophis macmahoni*, a similarly unusual species from western Pakistan, has peculiar, whorl-like rows of heavily keeled scales, with a shovel-shaped rostral scale. Many of these desert vipers have developed a unique sidewinding method of progression to overcome the difficulties of moving across loose, shifting sand (see p.28).

SUBFAMILY AZEMIOPINAE

There are only two species of *Azemiops* known to science, Fea's viper *Azemiops feae* and *A. kharini*. These vipers are rarely encountered, and occur only in the remote and wet mountain cloud forests of China, Myanmar, Vietnam and Laos. They are small- to medium-sized vipers, measuring up to approximately 1 m (3¼ ft). The scales on the top of the head are large and symmetrical, more like those of many other snakes such as elapids and colubrids than those of other vipers, which are typically smaller, more numerous and irregular. The body scales of *Azemiops* are all smooth, a feature they share with only one other species of viper, *Calloselasma rhodostoma*, a pit viper from Southeast Asia.

Little is known about the natural history of *Azemiops*, other than that they are ground-dwelling, oviparous snakes that lay clutches of approximately five eggs. Nocturnal or crepuscular, they feed on shrews and small rodents. Like pit vipers, they may vibrate their tails and gape when threatened, although their venom yields are very small and so they are not considered a threat to humans. When threatened, *Azemiops* will also flatten their bodies, including their head, which accentuates its triangular shape. The two species of *Azemiops* are superficially very similar and share a blackish body with narrow orange bars. They have somewhat different pale markings on the head and some detailed anatomical differences, though some researchers have suggested a more thorough analysis is needed of whether they are truly distinct species.

LEFT The Indochinese Fea's viper, *Azemiops feae*, is the only representative of the viperid subfamily Azemiopinae.

SUBFAMILY CROTALINAE: Rattlesnakes and Other Pit Vipers

The 280 species in the subfamily Crotalinae include the rattlesnakes (*Crotalus* and *Sistrurus*) and about 21 other genera distributed throughout the Americas and South and Southeast Asia. The most characteristic feature of the subfamily is a pair of infrared-sensitive pit organs on each side of the head used for locating prey. All pythons and at least some boas of the genera *Corallus*, *Epicrates* and *Boa* also have heat-sensitive pits, but these occur on, between or behind the scales of the lips (labials). In contrast, the crotaline vipers have a single large pit on each side of the face between the eye and nostril. The pit organs of the pit vipers are more sophisticated than that of boas and pythons; they consist of two compartments, divided in the middle by a membranous diaphragm. The smaller, inner part is connected by a narrow duct to a small pore in front of the eye, which appears to be a means of balancing the air pressure on either side of the diaphragm, and also measures the ambient air temperature. The larger outer chamber opens as a wide, forwardly directed aperture, through which infrared radiation emitted by prey enters and is detected by a series of highly sensitive thermoreceptive cells.

The infrared-sensitive pits of snakes are very sensitive to temperature variation, and experiments have shown that those of some pit vipers can detect changes in temperature of as little as 0.001°C (0.002°F). This renders these organs of use in

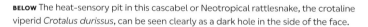

BELOW The heat-sensory pit in this cascabel or Neotropical rattlesnake, the crotaline viperid *Crotalus durissus*, can be seen clearly as a dark hole in the side of the face.

ABOVE The Central American jumping pit viper, *Metlapilcoatlus mexicanus*, photographed in Cusuco National Park, Honduras.

hunting even ectothermic prey such as frogs, which are often only marginally warmer than the surrounding environment. Using their pit organs, these vipers can locate prey, or predators, even in complete darkness. A pit viper deprived of its senses of sight and smell, for example, can perceive a mouse 10°C (18°F) warmer than its surroundings from a distance of 70 cm (28 in), and guide the direction of its strike accordingly to within about 5°. Pit vipers also select ambush sites based on either strong infrared contrast or thermal transitions, meaning that prey can more easily be detected.

Rattlesnakes, of which there are approximately 56 species, mostly in the genus *Crotalus*, are inhabitants of woodlands, prairies and rocky desert environments, mainly in North America and Mexico. One species, the cascabel, *C. durissus*, ranges from southern Central America far into South America. All species are rather stout-bodied snakes that range in length from the three 60 cm (24 in) long pygmy rattlesnakes (in the genus *Sistrurus*) to the big eastern diamondback, *C. adamanteus*, which may exceed 2 m (6½ ft). In newborn rattlesnakes the characteristic tail rattle, from which these pit vipers take their name, is at first only a small button. Additional segments accumulate with successive moults of

RIGHT Greatest of the pit vipers and largest of all venomous snakes in the western hemisphere, the bushmaster may reach an adult size of more than 3.5 m (11½ ft). Four species are recognized, of which this example, from Ecuador, is the common form, *Lachesis muta*.

the skin, and when the rattle becomes too large, the last and oldest segments break off. The rattle is used only for defence and at least one species, the Santa Catalina Island rattlesnake, *C. catalinensis*, has lost it during evolution; perhaps because it has no natural predators there is little need for such an adaptation, and, because it hunts arboreal prey, there is potentially a greater need for stealth.

None of the remaining pit viper genera has a rattle, although many will vibrate their tails among dry leaves when alarmed, which has the effect of creating a similar sound and is likely the evolutionary origin of rattlesnakes' rattles. Among the most diverse of the Latin American pit vipers are the more than 48 species of lanceheads (*Bothrops*), named after their characteristic lance-shaped heads. All are highly venomous, and they include such notorious species as the terciopelo, *B. asper*, from Mexico, Central America and northern South America, the South American jararaca, *B. jararaca*, and the exceptionally locally abundant golden lancehead, *B. insularis,* from Ilha da Queimada Grande in Brazil. So feared is the bite of the South American bushmaster, *Lachesis muta*, that in parts of its range it has been given the name matabuey, meaning 'ox-killer'. It is the world's largest viper, growing up to about 3.4 m (11 ft) with unconfirmed reports of specimens up to 4.5 m (15 ft). Bushmasters have rather peculiar knob-like scales, and they share this feature with Central and northern South American jumping vipers (*Atropoides* and *Metlapilcoatlus*), a group of seven exceptionally stout-bodied snakes named for their ability to strike with such force that the body is often carried forward with the momentum.

THE VIPERS OF QUEIMADA GRANDE

ABOVE Golden lancehead, the crotaline viperid *Bothropoides insularis*.

A small, deserted island off the eastern coast of Brazil barely 1 km (⅗ mile) across at its widest point, Queimada Grande is a refuge for large numbers of birds. It is renowned mostly, however, for its other main inhabitant, an endemic species of pit viper that occurs in extraordinary abundance. The golden lancehead, *Bothrops insularis*, is a slender, pale species that probably diverged from its larger mainland relative, the jararaca, *Bothrops jararaca*, when the island was separated from the Brazilian coast about 11,000 years ago. It subsists almost entirely on the various species of small birds that use the island as a reviving 'stop-off' point during migration, and relies on its potent, fast-acting venom to kill prey before they have a chance to fly away.

Walking through the island's wooded interior, it is not unusual to spot a golden lancehead every few paces. Commonly seen in trees, they may also be encountered among leaves on the ground, nestled between tree roots, on rock faces, and in almost every other accessible habitat. That so many of these snakes should occur in an area so small seems remarkable. There are, however, no other kinds of venomous snakes on Queimada Grande with which the lancehead might otherwise need to compete for food and shelter, and neither does it have any significant predators. With a more or less regular supply of birds, its principal food source is also almost always available.

Golden lanceheads are undoubtedly potentially dangerous snakes but, contrary to popular belief, their venom is in fact less lethal than that of many mainland *Bothrops* species, including their presumed nearest relative, *B. jararaca*, at least as assessed by mouse LD_{50} tests (see p.35). The erroneous reputation of high lethality in the venom of this species appears to have originated from some late-nineteenth-century experiments with unreported methods. More recent experiments have been unable to replicate results of high lethality of *B. insularis* venom. Some other pit vipers are similarly successful on other islands elsewhere, feeding on birds and attaining very high densities, notably *Gloydius shedaoensis* on Shedao off the northeastern coast of China.

RIGHT Caterpillar of the hawk moth, *Hemeroplanes triptolemus*, mimicking a neotropical pit viper. If molested, the caterpillar may even react with quick, sideways-directed 'strikes'.

Other ground-dwelling pit vipers from the Americas include eight species of the genus *Agkistrodon*. The eastern copperhead, *A. contortrix*, is an abundant species found over much of the eastern USA, and the cottonmouth, *A. piscivorous*, is very rare among vipers in being semi-aquatic. Nine hognosed pit vipers (*Porthidium*) are inhabitants mostly of neotropical lowland tropical forests, while two species of Mexican horned vipers (*Ophryacus*) and five montane pit vipers (*Cerrophidion*) live high in the mountains of southern Mexico and Central America. The 11 species of palm vipers (*Bothriechis*) are neotropical climbers, characterized by slender bodies, strongly prehensile tails, and beautifully mottled green colour patterns that are perhaps unsurpassed in concealing these snakes among leafy trees. One species, the eyelash viper, *B. schlegelii*, has impressive eyelash-like scales projecting from above the eyes that possibly aid in making the animal more cryptic or to help protect their eyes when moving through dense vegetation.

LEFT Indochinese white-lipped pit viper, *Trimeresurus albolabris*.

ABOVE Malayan pit viper, *Calloselasma rhodostoma*. This potentially dangerously venomous crotaline viperid is found throughout much of Southeast Asia, but despite its common name it occurs in only a small part of peninsular Malaysia.

Many Old World pit vipers belong to the ecologically diverse genus *Trimeresurus*, which has had a turbulent taxonomic history. These mostly arboreal snakes occur throughout much of Asia including numerous islands in the western Pacific Ocean. The Asian pit vipers are morphologically fairly conservative. Many species, such as Pope's pit viper *T. popeiorum*, the white-lipped pit viper, *T. albolabris*, and other members of the genus, display a similar and characteristic leaf-green colour pattern, sometimes only identifiable by species-specific hemipenes or DNA. Five similar Southeast Asian species now in the genus *Tropidolaemus* are distinguished from other species by keeled chin scales, and include the strikingly marked temple viper, *T. wagleri*. Six very thick-bodied species of *Ovophis* include the mountain pit viper (*O. monticola*) a species found throughout much of Asia, from India to China. One particularly large species, the Okinawa habu, *Protobothrops flavoviridis*, of the Ryukyu Islands in Japan, may attain lengths of 2.2 m (7 ft). This species has been exploited for use in habushu (snake wine).

Other Old World pit vipers include three species of hump-nosed vipers (*Hypnale*) from southwestern India and Sri Lanka and the Chinese sharp-nosed viper, *Deinagkistrodon acutus*, from southeastern China. These are ground-dwelling, tropical forest forms characterized by upturned snouts. Some 24 species of *Gloydius* are small inhabitants of temperate forests and mountains of Asia. One of these, *G. himalayanus*, occurs at elevations of up to 4,900 m (16,000 ft) in the Himalayas. The Malayan pit viper, *Calloselasma rhodostoma*, is widespread in Southeast Asia and the cause of many snakebite accidents in this region. In common with a number of other Asian pit vipers, and unlike most of its New World relatives, it is an egg-layer.

HOMALOPSIDAE: Mud Snakes

A highly aquatic family of snakes, members of the Homalopsidae tend to inhabit slow-moving or stagnant water, which gives rise to one of their common names, the mud snakes. From Southeast Asia and Oceania, they are found in both freshwater and coastal environments and have various specializations for aquatic life, such as nostrils set on the end of the snout that can be closed by valves when submerging, and small, upwardly facing eyes. There are currently 29 genera and approximately 57 species of homalopsids, a large number of genera for this

ABOVE The homalopsid, *Hypsiscopus plumbea.*

number of species. Mud snakes are rear-fanged but not regarded as dangerously venomous to humans. The majority of the living lineages of homalopsids evolved over the last 10 million years, though the eastern Indonesian genus *Brachyorrhos* diverged from the rest of homalopsids more than 40 million years ago. The five species of *Brachyorrhos* are rather different from other homalopsids in that they burrow in terrestrial soils, lack fangs and feed on earthworms. Most species are smallish snakes with relatively thick-set bodies, the stoutest being Bocourt's homalopsid, *Subsessor bocourti*, from West Malaysia, Cambodia, Thailand and Vietnam, and the longest probably being the masked water snake, *Homalopsis buccata*, with one female recorded as reaching a length of 1.37 m (4½ ft). Another Southeast Asian species inhabiting river mouths and coastal waters, the keel-bellied water snake, *Bitia hydroides*, has features more typically associated with some true (elapid) sea snakes, such as narrower ventral scales, and a somewhat flattened tail. Juvenile specimens even display a small head and narrow 'neck' followed by a large body, as seen in several true sea snakes (see p.147).

As far as is known, all homalopsids are viviparous, giving birth to baby snakes rather than laying eggs, although little is known about the reproductive ecology of several species. The nourishment of embryos via a placenta has been studied in

ABOVE The head of *Erpeton tentaculatum*, showing the sensory 'tentacles'.

some members of the family. Female homalopsids are known to continue feeding while gravid, unlike most snakes, which may reflect their ability to continue supplying nutrients to the embryos via a placenta.

Homalopsids feed mostly on fish and tend to eat a greater number of smaller prey items than most snakes, often consuming several in one feeding session, which they capture by ambush or by feeling around for them in muddy water. The tentacled snake, *Erpeton tentaculatum*, has long, paired protuberances on the end of its snout that have a sensory function, helping it detect prey by touch and vibrations. These tentacles are a few millimeters in length and extremely rich in nerve endings, enabling the snake to detect the slightest movement in the water, a feature that is particularly useful when foraging in murky environments. These completely aquatic snakes can remain underwater for up to 30 minutes without coming up for air. When hunting, they wait in a characteristic 'J' shape, with the bend between their body and head. Sometimes anchored to submerged twigs or plants by their tails, they strike to seize passing fish. Other homalopsids also use this technique, but the tentacled snake enhances its chance of catching a meal by exploiting its prey's escape reflex when hunting. When fish are disturbed, they dart quickly to the right or left, known as the 'C-start' reflex. The tentacled snake exploits this response by startling passing fish with a subtle movement that runs down the length of its body. Knowing the orientation of the fish, the snake can predict which way it will move and strike with startling accuracy. Often the fish turn directly towards the head of the snake, and sometimes even swim right into its mouth.

The five species of bockadams (*Cerberus*) are generally nocturnal in habit, and they inhabit estuaries and hunt mainly for mudskippers on mudflats exposed at low tide. The exception to this is the dog-faced water snake, *C. microlepis*, which occurs in only one freshwater lake (Lake Buhi) in the Philippines. Other mangrove-dwelling homalopsid species in the Indo-Pacific feed on crustaceans, including Gerard's homalopsid, *Gerarda prevostiana*, which eats freshly moulted mangrove crabs. This species has been observed wrapping its body around a crab and using its mouth to tear off manageable chunks, usually limbs. It is one of the few snakes known to break up its food before it eats. Another estuarine species to do this is the white-bellied mangrove snake, *Fordonia leucobalia*, distributed over much of the Australo-Papuan region, Southeast Asia and the Indian subcontinent. It is armed with robust fangs for piercing the hard exoskeletons of crustaceans. It eats mud lobsters and mangrove crabs in the intermoult (hard-shelled) stage and has been known to pull the legs off crabs that are otherwise too large to consume.

Paradoxically, crabs pose a considerable threat to small homalopsid snakes, as do large predatory fish and other snake species, including elapids such as the Indian cobra, *Naja naja,* and the banded krait, *Bungarus fasciatus.* For some homalopsids, such as the rainbow mud snake, *Enhydris enhydris,* common throughout Southeast Asia in large lakes disturbed by human activity, the most serious threat is human exploitation. In the early 2000s, the rainbow mud snake was found to make up about 70% of the catch in what was considered the world's largest snake-hunting operation. Every year at Tonlé Sap (the Great Lake) in Cambodia, nearly 7 million snakes of 11 different species were caught by local fishermen according to data collected between 2005 and 2006. Six of these species were homalopsids, used primarily as feed in the growing crocodile farm industry, although many *Homalopsis buccata* and *Subsessor boccourti* were sold for their skin. As stocks of fish became depleted in the lake, snakes offered a tempting alternative for use as human and animal feed and, since it was first documented in the late 1990s, the number of fishermen turning their attention to snakes increased dramatically. The sustainability of this practice or its long-term effects on the local snake populations are difficult to measure, but it seemed that catch per unit effort declined by about 80% during the period of the most-recent major study, which might indicate a substantial fall in numbers. Interestingly, *H. buccata* and *S. boccourti* populations seemed to show the greatest declines despite the overall catches for these species being quite low, which might be because these species produce relatively few offspring and because larger individuals were targeted for use in the skin trade, removing the most fecund females from populations. Detailed data have not been published for Tonlé Sap snake harvests since this study, so current trends are not known. A 2022 study of *E. enhydris* in West Java, Indonesia, concluded that this species remains locally abundant despite heavy exploitation by humans, perhaps due to their high fecundity and ability to exploit disturbed agricultural land.

THREATS TO SNAKES

Many snake species around the world are facing declines or are threatened with extinction. Habitat loss, over exploitation, disease (see p.24), invasive species (see p.74) and climate change all pose notable threats to snake populations. One measure that helps us to understand the stability of species and which ones are most vulnerable to extinction is the IUCN Red List of Threatened Species. Of the 3,509 snake species that have been assessed, over 12% are in a threatened category, with almost 20% listed as Data Deficient, meaning there is currently insufficient information to assign them a specific threat category.

One of the most significant threats to snake populations worldwide is habitat degradation due to modern human land-use patterns, both urban and agricultural. Other associated threats include the loss of prey species, particularly amphibians which have undergone catastrophic declines in recent years, due partly to diseases. Research has shown that in areas where amphibian species have declined, snake communities were generally less diverse and were also in decline.

Human consumption is also a significant factor contributing to the decline of some species. Globally, snakes are harvested for purposes such as food, leather products, medicines, the exotic pet trade, and other uses. Although some snakes are bred in captivity for these purposes, mostly wild-caught animals are consumed and traded, putting pressure on wild populations. Domestically the largest consumers of snakes are China and Southeast Asia. The largest single harvest of wild snakes occurs in the lake of Tonlé Sap in Cambodia, where millions of snakes are caught every year for

ABOVE Wild-caught water snakes, mostly homalopsids, on sale in a market in Cambodia.

human consumption as well as providing food for the crocodile farming industry (see p.122). The Convention on International Trade in Endangered Species of Wild Flora and Fauna (CITES) compiles data on the international trade of species listed within their Appendices. According to their database, 262,836 live snakes were traded in 2019, as well as 626,568 leather products, 582,062 skins and 278,688 derivatives. Several different *Python* species are the most commonly traded for their skins, with the oriental rat snake, *Ptyas mucosa*, the most traded live snake and Russell's viper, *Daboia russelii*, the most traded for derivatives.

ELAPOIDEA: Cobras and Relatives, and the Major Radiations of African Snakes

As explained on p.54, the previously very large and diverse family Colubridae as recognized in the late twentieth century was reorganized and subdivided, with the creation of several new family-level groups. In the previous, 2012 edition of this book, one of these was the Lamprophiidae, an Old World family with particularly strong representation in Africa. That concept of Lamprophiidae has since been shown *not* to represent a natural group because it comprises more than one evolutionary lineage, some of which are more closely related to the family Elapidae than to each other, and the major subfamilies it previously contained are also now generally considered to be distinct families. These family-level groups are now classified along with the family Elapidae (cobras, mambas, coral snakes and sea snakes) in the superfamily Elapoidea. The non-elapid Elapoidea includes approximately 340 species that (except for a handful of poorly understood species) are classified in the families Pseudoxyrhophiidae, Lamprophiidae, Prosymnidae, Psammophiidae, Pseudaspididae, Cyclocloridae, Atractaspididae, and Micrelapidae. The outliers to these families are the three species of African *Buhoma*, whose evolutionary relationships to other elapoids are not yet clearly understood.

FAMILY ELAPIDAE: Cobras, Coral Snakes, Kraits, Taipans and Sea Snakes

The Elapidae is a diverse family that includes some of the largest and most venomous snakes. Many elapids, such as the coral snakes, kraits and some sea snakes, are also among the most spectacularly coloured. As a group, the approximately 53 genera and 390 species are characterized by enlarged, non-erectile fangs in the front of the mouth (see p.21), which fit into grooved slots in the lower jaw when the mouth is closed, and they have a venom that is predominantly neurotoxic (see p.34).

Elapids also lack a loreal scale (on the side of the snout between the nostril and eye), a feature that in the superfamily Elapoidea otherwise characterizes only the stiletto snakes (Atractaspididae). Elapids occur in all the warmer regions of the world except Madagascar and several other island groups, and they are particularly well represented in Australia, where they reach their greatest diversity, and in terms of different species outnumber all other kinds of snakes.

Most terrestrial elapids are ground-dwellers, notable exceptions being the arboreal mambas (*Dendroaspis*) and tree cobras (*Pseudohaje*), and the aquatic water cobras (two species of *Naja*), all of which occur in Africa. Many elapid

ABOVE Red spitting cobra, *Naja pallida*, from northeastern Africa. When at rest or on the move, the hood of cobras lies collapsed along the sides of the neck as rather loose skin.

species are also burrowers, notably the coral snakes. The family also includes more than 70 marine species, the sea snakes and sea kraits, which, although different in habits and some features of anatomy, are nonetheless closely related. Their distinctive appearance is the result of adaptive modifications made necessary for living in the sea, rather than separate ancestry from terrestrial elapids.

Apart from one genus (*Acanthophis*, the death adders) of especially stout, viper-like terrestrial species, elapids are comparatively slender in build. Many have the bodily appearance of non-venomous colubrids, and in some, this similarity extends also to coloration. The red-yellow-and-black-banded coral snakes of the Americas, for example, bear striking resemblance to various 'mimic' species found in the same areas of distribution (see p.38 and p.45). Herpetologists have traditionally recognized several subfamilies of elapid snakes, but do not yet fully understand evolutionary patterns within this large and complex group and so many of the details of classification within the family are yet to stabilize. One of the only major groups within Elapidae that consistently appears in classification schemes is the subfamily Hydrophiinae, consisting of the Australasian terrestrial elapids and all marine elapids. Throughout the rest of this section we focus on several of the major groups within the Elapidae.

WHICH SNAKE IS THE MOST DANGEROUS?

ABOVE Asian saw-scaled viper, the viperine *Echis carinatus*.

The frequently asked question 'which snake in the world has the most lethal venom?' is difficult to answer decisively because much depends on the method of testing. However, the inland taipan or fierce snake, *Oxyuranus microlepidotus*, an Australian relative of the cobras, is perhaps more venomous than any other snake – a single bite delivers enough to kill more than 200,000 mice, or at least 12 adult humans. The eastern brown snake (the Australian elapid *Pseudonaja textilis*) is among the closest challengers. Whether or not the inland taipan is the most dangerous in terms of the everyday lives of humans, however, is a different matter. Inland taipans are shy and rarely encountered snakes found only in the remote outback of western Queensland, where few people have ever even seen one.

The group of species responsible for the greatest number of fatal accidents, and thus overall probably the most dangerous of all venomous snakes, are the saw-scaled or carpet vipers (*Echis*). These widely distributed snakes are found throughout much of northern Africa, the Middle East, India and Asia, often in close proximity to human habitation, and in places they can be remarkably abundant. Owing to their small size and highly cryptic colour pattern, they are difficult to detect and may easily be stepped on. They also have very potent venom, and can strike out with little provocation. Among various other potentially dangerous species, responsible for a large proportion of snakebite fatalities in humans, are the neotropical lancehead pit vipers (species of the genera *Bothrops* and *Bothrocophias*) from Central and South America; the puff adder (the viperid *Bitis arietans*) and mambas (the elapid *Dendroaspis*) of Africa; the cobras (the elapid *Naja*) of Africa, India and Southeast Asia; Russell's viper (*Daboia russelii*) of India and *D. siamensis* of Southeast Asia; and the elapids in the New Guinea and Australian region: death adders (*Acanthophis*), the common taipan (*Oxyuranus scutellatus*), the tiger snake (*Notechis scutatus*) and brown snakes (*Pseudonaja*). The World Health Organization has recently declared snakebite a 'Neglected Tropical Disease' (see p.101).

The symptoms and severity of snakebite for humans depend on a wide range of circumstances. including the size and age of the snake in question, quantity of venom injected, individual sensitivity of the victim, level of medical care and time taken to reach hospital. Even if one is unfortunate enough to be bitten by a venomous snake, this does not necessarily always lead to the injection of venom and the development of symptoms. Venom is a precious commodity that requires energy and time to produce and, as their principal means of obtaining food, most snakes will not expend it needlessly when defending themselves. Based on studies of proven bites by particular species, the ratio of bites to envenomation is usually about 2:1. It should also be remembered that, although potentially dangerous, many venomous snakes are extremely beneficial to humans by controlling pest species such as rodents.

Mambas

Among the most feared groups of elapids are the mambas (*Dendroaspis*), which occur only in Africa south of the Sahara. There are four species, of which the black mamba, *D. polylepis*, is infamous for its speed of movement (one of the fastest terrestrial snake species), unpredictable disposition and potent venom. Brownish in colour, it is mostly terrestrial, whereas the other species are predominantly green and live in trees. Mambas (from a Zulu word meaning 'big snake') are large, slender, agile snakes. The black mamba may attain lengths of up to 4.3 m (14 ft) and is extremely fast in its movements: across open ground it has been recorded at 15 km/h (9 mph), and on downhill slopes is almost certainly capable of moving even faster. The neurotoxic venom of black mambas is fast-acting and can be life-threatening within minutes. Just two drops can be a fatal dose in humans, and with each bite these snakes are capable of injecting up to 20 drops. Like cobras, they are able to flatten the neck, although not nearly to the extent of forming a distinct hood.

The black mamba occurs throughout the eastern half of tropical Africa, favouring dry, open bush country and living in abandoned animal holes, hollow trees or termite mounds where they may take up residence for several weeks or even months. They hunt mammals during the day, mostly bush babies, hyraxes,

LEFT Black mamba, *Dendroaspis polylepis*. The common name of this elapid snake alludes not to the colour of the body, which is uniform leaden-grey or olive brown, but to the purplish-black lining of the mouth.

gerbils and other small mammals, although they have also been observed to eat sugarbirds, catching them from the air as the birds hover around flowers, feeding on nectar. In pursuit of prey they may descend underground into rodent burrows and climb high into trees. Black mambas are only 38–61 cm (15–24 in) long when they hatch but grow rapidly: some may reach a length of 1.8 m (6 ft) before they are a year old.

Cobras and cobra-like elapids

Cobras and their relatives are distributed throughout much of Africa and Asia, with some growing to considerable lengths. At adult lengths of up to 5 m (16 ft), the hamadryad or king cobra, *Ophiophagus hannah*, is the world's largest venomous snake. Despite their common name, the king cobras are not true cobras, which instead comprise 34 species in the genus *Naja*. The forest cobra, *N. melanoleuca*, may occasionally exceed 3 m (10 ft), and several others reach lengths of 2.5 m (8 ft). A feature for which cobras are especially noted is their ability to spread the skin of their necks into a flattened hood. Several true cobras also have fangs specially modified for the purposes of 'spitting' venom (see p.131) along with

BELOW Although undoubtedly a potentially dangerous elapid snake, the Asian hamadryad or king cobra, *Ophiophagus hannah*, prefers to escape unless it is provoked. This is not true of nesting females, however, which may attack without provocation.

ABOVE A Moroccan specimen of the Egyptian cobra, *Naja haje*.

the closely related rinkhals, *Hemachatus haemachatus*, a viviparous species from Africa. All true cobras lay eggs, and the females of at least some species stay with their clutches throughout incubation. Female king cobras take particular care of their eggs by laying them in a nest that they build from leaves and small twigs, often deep inside bamboo thickets, and watching over them until they hatch. These nests contain two chambers, a lower one where up to 50 eggs are laid, and an upper one where the female often resides in order to protect the eggs.

Other African elapids include two species of aquatic cobras, restricted to the vicinity of large lakes. These were previously classified in their own genus (*Boulengerina*) but based on DNA evidence they do not seem to form a lineage entirely separate from 'typical' cobras, and so are now instead classified within the genus *Naja*. Their bodies are thick and heavy, and the neck can be flattened into a hood, although not to the extent seen in other true cobras. Much of their time is spent diving for fish, their principal food, and they may remain submerged for long periods at a time, but they are also to some extent terrestrial and when not in the water lie in crevices among rocks near the shoreline or bask in the sun. They grow to a length of about 2.5 m (8 ft). Two species of shield-nosed cobras (*Aspidelaps*) are 50–70 cm (20–28 in) long, stout-bodied snakes that have modified snouts for burrowing through soil and rooting out their prey, which are mainly frogs and lizards. Both species spread a narrow hood and hiss loudly when alarmed. Also from Africa are two species of tree cobra (*Pseudohaje*), large, slender snakes with very large eyes and long tails, an adaptation to climbing.

Two other genera of cobra-like snakes occur in the Afro-Asian region, and the relationships of these to other Old World elapids is less clear at present. The two desert black snakes, genus *Walterinnesia*, are 1 m (3¼ ft) long nocturnal species found in arid habitats of the Middle East, while the other genus (*Elapsoidea*) contains 10 mostly small, burrowing species of garter snakes in various parts of Africa (not to be confused with north American natricine colubrid garter snakes, see p.174).

Cobras were once classified into just a few, widespread species, as was the case with several other large and venomous snakes. More recent studies including the incorporation of DNA data have resulted in a very different understanding. Nowhere is this more evident than among the 'true' cobras of the genus *Naja*. In the early 1960s there were only six species recognized, when the previous edition of this book was published in 2012 there were 26 species recognized, and now there are considered to be 34 species.

ABOVE Defensive venom-'spitting' by the Mozambique spitting cobra, *Naja mossambica*.

Perhaps no other venomous snakes are more instantly recognizable than the cobras. With their head raised high and neck spread into a flattened 'hood', the appearance of one of these animals primed to defend itself is one of the most impressive spectacles in nature.

All cobras are capable of spreading a hood when threatened, which they achieve by extending the specially lengthened ribs of the neck. In addition, several species of cobra in Africa and Asia defend themselves by squirting venom through specially evolved fangs that have an orifice on the anterior surface rather than at the tip as in most other species. Venom is forced at high pressure through these fangs and directed at a potential predator in a well-aimed stream, usually at the eyes and face. In some African species the fangs even have spiral grooves inside that function much like the rifling of a gun barrel, helping to fine-tune the accuracy of the 'spit', as it has become misleadingly known. When 'spitting', cobras raise their neck well off the ground and tilt their head upwards and hold their jaws widely agape. By curling back the upper lip just enough to expose the orifice of the fang, however, some species can spit with the mouth only partially open, enabling them to perform this feat from almost any position. The effective range to which the venom can be ejected varies, but in the larger species, such as the black-necked spitting cobra, *Naja nigricollis*, it can be as much as 3 m (10 ft). Venom contacting the eyes causes pain and temporary or permanent loss of sight.

A link has been found between the evolution of spitting in cobras and the evolution of hominin primates – humans, chimpanzees, gorillas and their closest extinct relatives. The major lineages of spitting cobras in Asia and Africa both evolved at the time when hominins started appearing in those areas. Hominins spend more time standing and walking on their hindlimbs than most primates, and primates generally have an innate fear of snakes, so they might have used sticks and rocks to try and kill or scare snakes. This might have promoted the defensive spitting behaviour in cobras. Spitting cobras have additionally evolved venom (phospholipase A2 toxins and cytotoxins) that trigger sensory neurons of mammals to increase pain almost instantaneously. The independent evolution of spitting behaviour, anatomy and physiology in three different cobra groups supports the idea that spitting cobras evolved in response to the evolution and arrival of hominins.

New world coral snakes

Coral snakes are best known for their often vivid, red-yellow-and-black-banded colour patterns, which serve as warning signals that alert potential predators to their venomous character. Many animals that eat snakes, particularly birds of prey, appear to have an innate aversion to colourful markings and instinctively avoid them. There are currently two New World coral snake genera recognized, a single species of *Micruroides,* and 84 species of *Micrurus.* The numbers of scale rows around the body and the arrangement of scales on the head are almost constant, and their colour patterns tend also to be similar. Most species are restricted to tropical forests, although one species, the harlequin coral snake (*Micrurus fulvius*) occurs in the southern USA, and the Sechura coral snake (*M. tschudii*) and Sonoran coral snake (*Micruroides euryxanthus*) are desert species. They are typically slender snakes, with heads scarcely wider than their bodies, short tails and small eyes,

BELOW The elapid *Micrurus surinamensis* from tropical South America. Despite their potent venom, New World coral snakes often try to hide their heads and draw attention instead to their tails when disturbed. Many less venomous neotropical snakes have similar colour patterns and possibly mimic coral snakes occurring in the same region. Compare for example *Oxyrhopus rhombifer* (p.177).

and most are strikingly marked with a pattern of red, yellow and black rings. Exceptions include the white-banded coral snake, *Micrurus albicinctus*, from the lowland forests of Brazil, a predominantly black species with contrasting rings of white spots, and the Andean black-backed coral snake, *M. narducci*, characterized by a uniformly dark-coloured body and a red- or yellow-spotted abdomen.

New World coral snakes all have a specialized diet consisting mostly of other snakes, including their own species and other venomous snakes such as vipers, but also other elongate and limbless vertebrates such as amphisbaenian lizards and caecilian amphibians. Several species, including Allen's coral snake, *Micrurus alleni*, from southern Central America, eat swamp eels, and the diet of the Surinam coral snake, *M. surinamensis*, seems to be dominated by these and various other fish. Hemprich's coral snake, *M. hemprichii*, from northern South America is particularly unusual in seeming to specialize in eating onycophorans (velvet worms). Coral snakes are all active foragers, and though many appear to be generally nocturnal, others have no particular set pattern of activity.

The bites of New World coral snakes can be dangerous. It is perhaps only the small Sonoran coral snake, *Micruroides euryxanthus*, that produces little more than 6 mg of a relatively weak venom that is not potentially lethal to humans. New World coral snakes have rather short fangs and their mouths are relatively small, which has given rise to the myth that these snakes are incapable of biting humans unless they happen to fasten on to a thin piece of skin, such as that between the fingers. Actually, even the smallest species have a surprisingly wide gape and are able to deliver a bite to almost any part of the body. Although armed with powerful venom and strong warning colours, several species also twitch their whole bodies rapidly in curious, often unnerving, jerking movements that can make it difficult to decide which is the head and which the tail end, providing additional discouragement to potential attackers.

Asian coral snakes and kraits

Asian coral snakes appear to be closely related to the New World forms and, like those species, many have vivid colour patterns. Two species from Southeast Asia are particularly brightly coloured: the blue Malayan long-glanded coral snake, *Calliophis bivirgatus*, and the banded coral snake, *C. intestinalis*, which is brownish with a red or orange dorsal stripe enclosed between two black stripes. The venom glands of these species are especially large, extending under the skin for about one-third the length of the body. In common with most New World coral snakes, they lay eggs and feed on other snakes. Approximately 13 other coral-snake-like species grouped in the genus *Calliophis* occur over much of Asia, including India and Sri Lanka, southern China, Japan and the Philippines. These are small semi-fossorial species with exceptionally slender bodies and small heads and eyes. They are nocturnal and feed mostly on other reptiles, especially snakes.

Similar to the Asian coral snakes in appearance, although somewhat larger, are the 16 species of kraits (*Bungarus*). Kraits occur over much the same range as Asian coral snakes, and like them also feed chiefly on other snakes. An exception is the many-banded krait, *B. multicinctus*, which mainly eats fish. Most have an enlarged mid-dorsal row of scales and peculiar protrusions on the vertebrae, the precise function of which has not been established but may play some defensive role in body thrashing, to which these snakes often resort when molested. As with coral snakes, some are brightly coloured. Kraits are highly venomous and are responsible for human fatalities throughout their range. Many bites occur when the unfortunate person is sleeping on the floor and they roll over onto the snake and, due to the short fangs in this genus, many people do not notice the bite as soon as it happens.

ABOVE Malayan long-glanded or blue coral snake, *Calliophis bivirgatus*. This spectacularly coloured elapid occurs throughout much of southeast Asia. It is nocturnal and secretive.

ABOVE Banded krait, the Asian elapid *Bungarus fasciatus*. This species is among the largest of kraits, with adults reaching lengths over 2 m (6½ ft).

Terrestrial elapids of the Australo-Papuan region

The greatest numbers of terrestrial elapid snake species are found in Australia and New Guinea (the Australo-Papuan region), where in places they are the dominant snake species. In mainland Australia alone there are 108 different species, compared to only 79 species across all four other families of snakes that occur there. Almost all terrestrial Australo-Papuan elapids are ground-dwelling or at least partly burrowing, and of the few species that occasionally climb, none shows any clear specializations for arboreal life. Quite why there should be so few elapids in the Australo-Papuan region that live in trees, where this habitat is otherwise exploited only by a few pythons and colubrids, is puzzling, but it is interesting that their nearest relatives in Southeast Asia are also mostly ground-dwelling. Perhaps a suitable opportunity for diversification among these forms has never arisen or their ancestors had little natural inclination, need or ability to climb, but this does not seem to be the complete story, especially when it is considered how many other ecological niches in this region are now occupied by elapids.

ABOVE One of the most formidable of all Australian-Papuan elapids, the taipan *Oxyuranus scutellatus,* is also the largest venomous snake in this region. There are authenticated records of examples up to 4 m (13 ft). This example is from Papua New Guinea.

Australo-Papuan terrestrial elapids do not form a natural group because their last common ancestor also gave rise to the two separate groups of marine elapids, the sea kraits and sea snakes. Thus, terrestrial and marine Australo-Papuan elapids together comprise a natural group, the subfamily Hydrophiinae. This is a prolific radiation of venomous snakes that evolved surprisingly rapidly, probably within the last 14 million years or so according to estimates from rates of DNA evolution, as calibrated from the fossil record.

Taipans, brown snakes and whip snakes

The terrestrial elapids of Australia and the New Guinea region comprise two principal types, one of which includes mostly egg-laying species with a paired row of scales beneath the tail, and the other of which is viviparous with a single row of subcaudal scales. Notable among the larger egg-laying forms, three species of taipans (*Oxyuranus*) are arguably the most fast-moving, unpredictable, and venomous of all Australo-Papuan elapids. Drop-for-drop, the venom of the inland taipan, *O. microlepidotus*, is perhaps more potent than

SNAKE EGGS IN ANT NESTS

BELOW Short-tailed coral snake, the South American elapid *Micrurus frontalis*, with tail raised in typical defence display.

Coral snakes usually lay their eggs in leaf litter, beneath rotten logs, or in other places similarly conducive to incubation. The short-tailed coral snake, *Micrurus frontalis*, in South America, however, has found a particularly novel way of taking care of its eggs and ensuring that they are provided with optimum conditions. It lays its clutch of one to seven eggs in the nest mound of a particular species of ant, *Acromyrmex lobicornis* – specifically in that part of the nest used by the ants for cultivating a fungus on which they feed. In this underground chamber, the humidity remains more or less constant, and the temperature varies by little more than 2°C (4°f), providing an ideal incubation environment for the developing eggs. The ants also clean the eggs, reducing the risk of them becoming contaminated by

bacteria or mould, and will even protect them from attack by predatory insects. It is not entirely clear how or indeed if the ants benefit reciprocally from this behaviour, but it may be that the newly hatched coral snakes provide some protection by feeding on amphisbaenians (burrowing, legless, snake-like lizards) and scolecophidian snakes, which are natural predators of ants and often invade their nests. However, some scolecophidians, such as the blind snake *Liotyphlops albirostris* also lay eggs in fungus-growing ant nests.

Other snakes are known to use anthills and also termite nests for incubating their eggs. Among these, the Patagonian green snake (the dipsadine colubrid *Philodryas patagoniensis*) may lay its eggs in the same ant nests as those used by short-tailed coral snakes.

that of any other snake in the world (see p.126). This species is associated primarily with the flat plains of central Australia so rarely comes into contact with humans. Closely related to the taipans are nine species each of the black snakes (*Pseudechis*) and brown snakes (*Pseudonaja*), named in allusion to their predominant body colour. Among the most widespread, both in Australia and possibly also New Guinea, is the mulga or king brown snake (although belonging to the 'black snake' genus), *Pseudechis australis*, a formidable 2.5 m (8 ft) long species that is often unperturbed in the presence of humans, and reluctant to move away when encountered. A high degree of DNA variation has been found in *P. australis* across its wide range.

All black and brown snakes are potentially dangerous and, if provoked, some are highly aggressive. A large New Guinea species, the Papuan black snake, *P. papuanus*, is reputed to attack with a tenacity unrivalled by other species, a reputation which has earned it the local name of auguma (meaning 'to bite again'). However, within New Guinea many people call most or all snakes 'black snakes' when in fact most people will be referring to the more commonly encountered New Guinea taipan, *Oxyuranus scutellatus canni*. Brown snakes and black snakes feed on frogs, lizards, small mammals and occasionally birds, and like many elapids in this region, often constrict their prey as well as injecting venom. With long slender bodies, 14 species of whip snakes (*Demansia*) are among the most agile of all Australo-Papuan elapids. Though normally day-active, they may also be active at night when the weather is warm. Although all are venomous, only large specimens are regarded as potentially dangerous to humans.

Tiger snakes, death adders and other viviparous species

Viviparity is uncommon among terrestrial elapids. In Africa and Asia, only the rinkhals, *Hemachatus haemachatus*, and a few other species produce offspring by this means, and in tropical America elapids are exclusively oviparous. In Australian elapids, however, viviparity is more prevalent, especially among species found in the cooler, southern part of the continent, and some species have relatively large numbers of offspring. The litters of tiger snakes may contain more than 40, and a female black tiger snake, *Notechis ater*, from Tasmania was observed to give birth to 109 babies, more than has been recorded in any other Australian snake. Tiger snakes are large, up to 2.5 m (8 ft) long, stocky species widely distributed in southern Australia, and black tiger snakes also occur on many of the offshore islands, often in high densities. Males appear

ABOVE Tiger snake, the Australian elapid *Notechis scutatus*.

to be much stronger than females of the same body size, perhaps because of the strength needed during bouts of ritual male-to-male combat, or because females need to store more fat for reproduction. Also among the largest of Australia's live-bearing elapids are three species of copperhead (*Austrelaps*). Like the tiger snakes, they are more resistant to cold than most other species and can sometimes be found sunbathing even in winter.

A group of eight viviparous species from Australia, New Guinea and nearby islands that are extremely well-camouflaged and have venoms of exceptional potency, are the death adders (*Acanthophis*). Whereas virtually all other elapids are rather slender snakes that actively hunt for their prey, these are sedentary, particularly heavy-set snakes that lie in wait for small animals to pass within striking distance, often using the tip of their brightly coloured tail as a lure for small birds or mammals. As a consequence, they are often difficult to see, and in places represent a serious potential risk – fatal accidents attributed to

ABOVE Common death adder, *Acanthophis antarcticus*. Death adders are elapids that resemble and behave like vipers. In their native Australia and New Guinea, where there are no vipers, these snakes occupy the same ecological role.

envenomation by these snakes are reported each year in New Guinea. Their home ranges may be smaller than those of any other elapids, and within an area of only a few square metres they may not move around very much for weeks at a time. Another small adder-like elapid, the bardick, *Echiopsis curta*, is an exclusively Australian species that feeds mostly on frogs and also uses a 'sit-and-wait' hunting technique.

Three species of broad-headed snakes (*Hoplocephalus*) are slender, nocturnal and the only elapids in Australia that are frequently arboreal. They feed principally on lizards, but occasionally eat frogs and mammals. A jet-black snake dotted with yellow scales, *H. bungaroides* is a particularly spectacular species confined mostly to rocky sandstone habitats from southeastern districts, where its survival is under threat from commercial 'bushrock' collectors. Now classed as Vulnerable, at the time of European settlement during the nineteenth century this snake was common even in the centre of Sydney.

*Australian coral snakes, shovel-nosed snakes, crowned snakes
and forest snakes*

Several genera of small, mostly egg-laying elapids in Australia and New Guinea are noted for their burrowing habits, defensive displays, and unusual diets. These include four species of Australian coral snakes (*Simoselaps*) and eight species of shovel-nosed snakes (*Brachyurophis*), which are variously marked with red-and-black or yellow-and-black bands and have a specialized diet consisting almost wholly of reptile eggs. Unlike those of other elapids, the teeth on the pterygoid bones in at least one species, the narrow-banded shovel-nosed snake (*B. fasciolatus*), are saw-like. Although sharing the same common name, these Australasian 'coral snakes' are not part of the same lineage as Asian and New World coral snakes. The black-and-white-ringed bandy-bandy, *Vermicella annulata*, is a similarly patterned Australian elapid and one of six species in this genus that eat mostly blind snakes (typhlopid scolecophidians). When alarmed, it assumes a curious defensive posture in which the body is elevated in large loops and twisted around, probably as a means of increasing the effectiveness of its black-and-white warning bands (see p.48). The four species of Australian crowned snakes (*Cacophis*) and perhaps the three species of New Guinea crowned snakes (*Aspidomorphus*) also adopt unusual defensive postures if provoked. The dorsal coloration of these snakes is generally brownish, but the head and neck are more boldly marked with a white or yellowish stripe. When danger threatens, these snakes arch their forebodies off the ground but keep the head pointing

ABOVE Australian banded snake,
the elapid *Simoselaps littoralis*.

directly downwards, thus displaying their colourful head markings to their full effect. Twenty-four species of forest snakes (*Toxicocalamus*), all of which occur only in New Guinea, are among the very few elapids that feed on invertebrates. Earthworms comprise a large proportion of their diets, but they may also eat insect larvae and pupae, and small snails. They are small, secretive snakes with many species being fossorial or semi-fossorial, spending much of their time in leaf litter and soil or under rocks and logs.

Small-eyed snakes, Indonesian coral snakes and the Fijian bola

Although relatively close to Australia in geographical terms, New Guinea, the Solomon Islands and Fiji have several distinctive elapid snakes that occur only in these areas. Most of these are monotypic genera (containing only one species each). During the day, the New Guinea small-eyed snake, *Micropechis ikaheka*, shelters beneath leaf litter and other forest floor debris, or beneath the discarded heaps of coconut husks in plantation areas, where it poses a potential danger to local workers. This generally banded species can have a pale coloration throughout most of its range, a feature for which it has become known in parts of its native land as 'white snake'.

ABOVE The small-eyed snake, *Micropechis ikaheka*, is a fairly large – up to 1.5 m (5 ft) long – elapid endemic to New Guinea and adjacent islands, where it is a nocturnal inhabitant of monsoon forests and also occurs in coconut plantations.

Similar to the small-eyed snake is *Loveridgelaps elapoides* from the Solomon Islands, a strikingly marked black-and-white-banded species with patches of bright yellow on its back. This rarely seen nocturnal elapid occurs mostly near forest streams and feeds on frogs, a habit it shares with another endemic, diurnal species, *Salomonelaps par*. From the island of Bougainville near New Guinea, Hediger's coral snake, *Parapistocalamus hedigeri*, is a small elapid, up to 50 cm (20 in) long. Very little is known about it other than it is a nocturnal, semi-fossorial species that may feed on the eggs of large land snails. Perhaps an even more poorly understood species is the Fijian ground snake, or bola, *Ogmodon vitianus*. This snake has one of the most isolated distributions of any elapid in the region. It occurs only on the small island of Viti Levu, nearly 2,000 km (1,240 miles) from its nearest relatives in the Solomons. Its known distribution appears to be further limited to two adjacent watersheds in the southeastern part of this island. A diminutive, burrowing species, only about 20 cm (8 in) long, it is found in forest soils of inland mountain valleys and eats mainly earthworms.

Marine elapids – the sea snakes and sea kraits

Among the most intriguing of all snakes, in terms of their origins, relationships and specialized life habits, are the sea kraits and sea snakes. The 72 or so species include some of the most completely aquatic of all air-breathing vertebrates, and the only living reptiles that spend all of their lives at sea. With a few exceptions, sea kraits and sea snakes are exclusively marine. Among sea snakes, *Hydrophis semperi* is endemic to Lake Taal in the Philippine Islands, and several species are known to inhabit rivers in Asia and the Australo-Papuan region. Among sea kraits, *Laticauda colubrina* and *L. crockeri* inhabit a brackish lagoon (Lake Te Nggano) on Rennell Island in the Solomons, though the former species also occurs more widely. The two *Laticauda* species in Lake Te Nggano appear to avoid competition by feeding on different prey, the larger *L. colubrina* on eels and the smaller *L. crockeri* on gobies. Herpetologists once thought that all marine elapids (sea kraits and sea snakes) represented a single, divergent line of evolution, but the weight of morphological and DNA evidence now available indicates conclusively that there are at least two principal groups, the sea kraits and the true sea snakes. These are both lineages within the Elapidae, but are not each other's closest relative, and they evolved marine habits independently from different terrestrial ancestors within the Australo-Papuan region.

The sea kraits (Laticaudini) are the sister lineage to all other, terrestrial and marine, Australasian elapids (Oxyuraninae), and this whole group is together termed the Hydrophiinae. Sea kraits drink freshwater, and come ashore to breed and lay eggs. The 'true' sea snakes (Hydrophiini) are deeply nested within the evolutionary tree of Oxyuraninae, and therefore clearly evolved from terrestrial or semi-aquatic Australo-Papuan elapids. True sea snakes are more highly specialized for a marine existence than sea kraits and they occupy a much wider range of habitats. They do not lay eggs but instead are viviparous (give birth), and though some semiaquatic mangrove species may move around on mud between burrows at low tide, and a few more fully aquatic species may occasionally be seen on rocks or exposed reefs, most never leave the water.

Marine elapids are found in most tropical seas but are absent from the Caribbean and the Atlantic Ocean. While they extend to the far western Indian Ocean and into the Gulf of Oman, as well as into Costa Rica's Gulf of Dulce, they are mostly creatures of Asian and Australian coastal waters. Only a few species range far out to sea. In fact, only one species, the yellow-bellied sea snake, *Hydrophis platurus*, is truly ocean-going. At times, people have seen large aggregations of some sea snake species floating on the surface of the sea. In 1932, passengers aboard a steamer passing through the Strait of Malacca off Malaysia reported seeing a phenomenal number of sea snakes massed together in an enormous 'slick', which they said was about 3 m (10 ft) wide and extended for a distance of some 96 km (60 miles). However, there are no photographs of that event or contemporary observations of anything like this on such a vast scale, so it is difficult to know what relevance it has to our understanding of sea snake biology.

Sea kraits

All eight species of sea kraits are included within a single genus, *Laticauda*. As egg-layers, they are more tied to land than are the viviparous sea snakes. During the reproductive season they come ashore at night, often in large numbers, on beaches, or in wooded areas at the junction of water and land, with females often depositing their eggs communally in caves. At other times, especially after feeding, sea kraits may also crawl out onto logs floating in the sea or emergent rocks near the tide line to sun themselves. They are good climbers and can sometimes be seen under jetties. Most *Laticauda* are strongly associated with coral reefs and they are familiar to many snorkelers and scuba divers. They have broad ventral (belly) scales, legacies of their terrestrial ancestors, blue or yellow colour patterns banded with black, and reach a maximum size of about 2 m (6½ ft). Most feed mainly on eels, which they locate by poking their heads into holes or crevices. As with many aquatic snakes, females are conspicuously larger than males. Females

ABOVE Banded or yellow-lipped sea krait, *Laticauda colubrina*, swimming over a coral reef beneath a large shoal of fish, Malaysia.

may also feed in shallower water and prey on different kinds of eels than their mates. In parts of the Philippines, sea kraits used to be very heavily exploited, especially for the leather industry. However, most of these *Laticauda* 'fisheries' have collapsed. The remaining harvest is local and smaller in scale, such as on the islands of Cebu and Luzon, and largely supplies smoked meat to Japan. This same market is also supplied with *Laticauda* collected in the Japanese Ryuku islands.

True sea snakes

Although superficially similar in overall appearance, the true sea snakes (Hydrophiini) are more specialized for ocean life than the sea kraits. The six genera and approximately 64 species are all viviparous and spend their entire lives at sea, although three Australian species at least are more associated with mudflats. In length, they range from as little as 50 cm (20 in) in the few semiaquatic species and 85 cm (33 in) in some species of *Aipysurus* to 2.75 m (9 ft) in the yellow sea snake, *Hydrophis spiralis*. The most massively built is Stokes' sea snake,

H. stokesii, which at an adult length of 2 m (6½ ft) may have a midbody girth of over 25 cm (10 in) and total mass in excess of 2 kg (4 lb ½ oz). The three species of the semiaquatic Australian genera *Ephalophis*, *Hydrelaps* and *Parahydrophis* are restricted mostly to estuaries and mudflats, and they retain several relatively primitive features from their terrestrial ancestors, such as broad ventral scales. At the other end of the spectrum, the yellow-bellied sea snake, *Hydrophis platurus*, is highly adapted to oceanic environments, with a geographic range greater than that of any other snake or lizard. This species, uniquely marked with a pattern of yellow and black stripes, is found across the Pacific as far west as the western coasts of Central and South America, and as far south as New Zealand and Africa's Cape of Good Hope. It occurs mainly in the narrow strips of calm water where two ocean currents meet, feeding on small fish that congregate around the accumulations of seaweed debris often found floating in these areas.

BELOW The hydrophiine elapid *Aipysurus duboisii*, a large species of reef sea snake from the Timor Sea and coastal regions of northern and western Australia.

Adaptive modifications of marine elapids

- The body is flattened from side to side, with a longitudinal keel along the belly that probably acts as a keel-like stabilizer.
- All *Hydrophis* species have lost the tight one-to-one association between numbers of vertebrae and numbers of ventral scales, and the broad ventral scales of their terrestrial ancestors have been reduced in width.
- Some species have unusually small heads and narrow necks in relation to their otherwise stout bodies, which enables them to reach deep into holes to seize burrowing eels.
- All species have flattened, paddle-like tails with which to propel the body when swimming. In species of *Hydrophis*, this structure is supported by elongated neural spines (projecting up from the vertebrae), and in *Aipysurus* it is supported by haemapophyses (projecting down from the vertebrae) on the vertebral column of the tail. The difference in supporting structure in *Hydrophis* and *Aipysurus* perhaps suggests that the elaboration of a tail paddle evolved independently in these two groups, from an ancestor that had a less well-developed tail paddle as in the more semiaquatic species of Hydrophiini.
- Some species of sea snakes in the genus *Aipysurus* have photoreceptors (light-sensitive cells) on the tail, which they use to ensure that this end of the body is not left exposed when hiding among crevices during the day.
- Although a very small left lung is retained in a few species, most have a single lung that is longer than that of most other snakes (enabling them to stay underwater longer). Stored air can be pumped forward to the part with a richer blood supply to sustain respiration. The trachea (windpipe) is also modified to be capable of respiration. Sea snakes can also absorb oxygen through their skin to some degree, and at least one species (*Hydrophis cyanocinctus*) has an elaborate network of blood vessels and sinuses below the skin of the top of the head that likely serves to oxygenate the brain during extended periods underwater.
- Most species are active at depths of less than about 30 m (100 ft), although some species can forage in the dark, cold 'twilight' zone at depths of 250 m (820 ft). They can stay underwater for at least two hours.
- The nostrils are equipped with valves to keep out seawater. In sea kraits (*Laticauda*) the nostrils are placed on the side of the snout, while in sea snakes they are on the top.
- A modified rostral scale in *Hydrophis* (or an extension of tissue behind this scale in *Aipysurus*) fits into a notch at the front of the lower jaw and seals the opening through which the tongue is normally protruded while the snake is underwater.

ABOVE Elegant sea snake, the hydrophiine elapid *Hydrophis elegans*, a species from coastal regions of western New Guinea and northern Australia. Note the paddle-like tail adapted for swimming.

- A special salt excretion gland beneath the tongue enables marine elapids to rid themselves of excessive salt. The skin of these snakes is also more impermeable to salt than that of their terrestrial relatives.
- Sea kraits and sea snakes shed their skin more frequently than do terrestrial species, at intervals of 2–6 weeks. This may have some effect in helping to keep the body free of algae, barnacles and other small marine organisms.
- Adaptation to spectral sensitivity of longer wavelengths aids sea snakes to see underwater. Some species are capable of shutting their pupil down to a tiny pinhole when exposed to the much brighter light at the sea surface.

The venom of most marine elapids, although produced only in relatively small amounts, is powerfully neurotoxic (see p.34). Sea krait and sea snake venom is simpler than that of their terrestrial relatives. This streamlining is an adaptation to their fish diet, and it has evolved convergently in a very similar way in these two independent lineages, to such a degree that the same antivenom can work in cases of envenomation from both groups. Most sea krait species tend to be docile and may even be handled (not something we recommend) without being bitten, whereas some sea snakes bite more readily. The beaked sea snake, *Hydrophis schistosus*, in particular appears to be more aggressive than most and is suspected of being responsible for deaths among Malaysian fishermen.

Fish form the staple diet of most marine elapids, and some species specialize in feeding on particular types of fish. Sea kraits (*Laticauda*) hunt for eels among coral reefs, including potentially dangerous moray eels, which they usually release following the initial strike and leave for the venom to take effect before attempting to consume. Beaked sea snakes, *Hydrophis schistosus*, forage along the muddy bottom of estuaries for toxic catfish and even some of the most notoriously toxic pufferfish. The diet of some species, such as the olive sea snake *Aipysurus laevis*, also includes crabs and other crustaceans. The white-spotted sea snake (*Aipysurus eydouxi*), mosaic sea snake (*A. mosaicus*) and at least two of the three species of turtle-headed sea snakes (*Emydocephalus annulatus* and *E. orarius*), scrape fish eggs off rocks using enlarged anterior labial scales. These egg-eaters have a somewhat degenerate venom apparatus and their venom is also relatively weak, an adaptation to a shift away from catching live adult fish.

Sea snakes tend to produce smaller clutches and litters than do their terrestrial relatives. They also tend to produce larger newborns, probably because these need to be fully able to swim from birth. Some species, such as *Hydrophis viperinus*, produce small litters of three to four relatively large young, whereas others give birth to larger numbers of slightly smaller offspring.

FAMILY ATRACTASPIDIDAE: Stiletto Snakes and Centipede-eaters

Atractaspididae comprises about 70 living species that began diversifying approximately 40 million years ago. At least some atractaspidids were previously considered by some biologists to be strange vipers, but both morphological and DNA evidence strongly indicates that they are two separate groups (subfamilies Atractaspidinae and Aparallactinae) that are more closely related to elapids and colubrids than to vipers. Although atractaspidines and aparallactines are generally considered distinct subfamilies (as they are in this book), there is not yet compelling morphological or DNA evidence that they are both distinct evolutionary lineages. For example, some DNA evidence suggests that *Atractaspis* is more closely related to *Xenocalamus* (typically considered an aparallactine) than it is to *Homoroselaps* (typically considered an atractaspidine), so some future changes to classification might occur within the family Atractaspididae.

Aparallactines have fixed, grooved fangs in the rear of the mouth, resembling the rear-fanged (opisthoglyphous) colubrids and non-elapid elapoids. In contrast, atractaspidines have hollow, front-mounted fangs superficially like those of

RIGHT Duerden's stiletto snake the atractaspidine atractaspidid, *Atractraspis duerdeni*, from southern Africa. Stiletto snakes are found throughout much of Africa, with some species ranging into the drier northern part of the continent and also across large areas of Arabia.

vipers (see p.21). Atractaspidids are mostly adapted for burrowing. Their bodies are cylindrical and of about the same circumference throughout, with little or no discernible narrowing at the 'neck'. The skull is compact, and the small head often has a projecting snout. Many species have small eyes. The tail is typically very short, and in some species bears a sharp spine at its tip. Some of the larger species grow to just over 1 m (3¼ ft), though most are considerably smaller.

Several atractaspidines and aparallactines are potentially dangerous enough to be considered medically important. The bite of the Natal black snake, *Macrelaps microlepidotus*, in particular, has been known to cause a temporary loss of consciousness, and bites from the larger species of *Atractaspis* may have serious consequences. The venom is predominantly neurotoxic in its effect (see p.34), although it also produces local swelling, severe pain and other symptoms more typical of viper bites.

The atractaspidines and aparallactines are essentially African in distribution. Aparallactines and most atractaspidines occur in sub-Saharan Africa, but some atractaspidines range also into the Near East. Atractaspiids occur in habitats as diverse as rainforest, grasslands and semi-desert. Except for one species, Jackson's centipede-eater, *Aparallactus jacksonii*, they are all viviparous.

Subfamily Atractaspidinae: Stiletto snakes

The atractaspidines comprise 24 species, with all but two species of *Homoroselaps* classified within the genus *Atractaspis*. The stiletto snakes, also known as 'burrowing asps' or 'mole vipers', are remarkable for their disproportionately large, hollow fangs, which can be erected independently of each other and extended

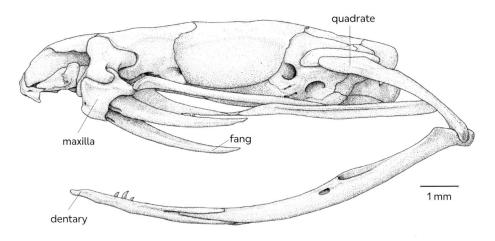

ABOVE Skull of the African stiletto snake, the atractaspidid, *Atractaspis aterrima*. Note the large backward-facing fangs on the upper jaw. On the basis of their dentition, *Atractaspis* species were long believed to be vipers. Unlike vipers, however, the fang-bearing maxillary bone pivots on a lateral ball and socket joint, and the fangs cannot be rotated far forwards.

downwards and slightly outwards into a biting position without opening the jaws. There are typically two fangs on each side, one functional and the other a replacement. All other maxillary teeth have been lost, and except for a few on the palatine bones and two or three on the dentaries, the mouth has no other teeth. The venom-injecting apparatus of *Atractaspis* is unusual in a number of other respects, too. Each fang bears a small cutting edge opposite its orifice, and the venom gland in some species, such as the small-scaled stiletto snake, *A. microlepidota*, is extraordinarily long, extending under the skin behind the head for approximately 30% of the body length. The venom itself also has a special composition (see p.36).

Stiletto snakes hunt and feed underground, on rodents and their nestlings but also on caecilian amphibians and lizards. Skinks, amphisbaenians (worm-lizards), frogs, and other snakes are also eaten. Without the need to open the mouth to bite, stiletto snakes are capable of killing prey in the most restricted of spaces, including underground burrows. In delivering the predatory strike, a single fang is erected from whichever side of the head is next to the animal, and jerked sideways, downwards and backwards with a quick 'stabbing' movement. The rotating maxilla displaces the upper lip, thus opening a slit through which the fang can be extended. Should a foraging snake encounter more than one rodent at a time, it will typically bite and immobilize all available prey before beginning to feed, and it may consume the occupants of an entire nest of mice in this manner if presented with the opportunity. In response to an assault from a predator, stiletto snakes will arch the neck and strike rapidly with a slashing backwards movement. They bite with little provocation, and owing to their peculiar fang erection mechanism, narrow heads, and the unusual flexibility of the neck vertebrae, are very difficult to restrain safely if handled. If molested, an *Atractaspis* may also use the tail-distraction ruse or tail poking (see p.47, p.49).

Subfamily Aparallactinae: Centipede-eaters, snake-eaters and their allies

Aparallactine snakes typically have one or two enlarged, grooved fangs that are situated towards the rear of the mouth and preceded, but not followed, by three to ten smaller, ungrooved teeth. Unlike those of stiletto snakes, the fangs are non-erectile. One species of centipede-eater, *Aparallactus modestus*, differs in having uniformly sized, ungrooved maxillary teeth. This subfamily includes eight genera and 44 species. The monotypic Natal black snake, *Macrelaps microlepidotus*, an inhabitant of damp places near water, has the least specialized diet of the aparallactines, eating rodents, frogs, legless lizards and a variety of other small vertebrates. The remaining seven genera have markedly diverse feeding habits.

The 11 species of *Aparallactus* have one of the most unusual dietary habits of any snake. They feed almost exclusively on centipedes, and have enlarged anterior mandibular teeth that perhaps enable them to grasp more effectively the hard, chitinous exoskeletons of these formidable invertebrates. Once bitten, a struggle often ensues until the envenomated centipede, which may be over 12 cm (5 in) long and nearly twice the diameter of the snake's body, is sufficiently disabled for the snake to swallow it. The bites of centipedes, which are themselves venomous, appear to have little effect on these snakes. *Aparallactus* occur in rainforest, open bush, sandy regions and savanna, where they are found among roots, beneath stones or fallen logs, and in termite mounds. In general appearance many species superficially resemble snakes of the New World colubrid genera *Tantilla* and *Tantillita*, which are also rear-fanged and at least some of which feed on centipedes.

Nine species of purple-glossed snakes (*Amblyodipsas*) – the common name refers to the purplish iridescent sheen of their mostly dark-coloured bodies – are rather stout-bodied snakes in which females appear to grow larger than males. At an adult length of 1 m (3¼ ft), female common purple-glossed snakes, *A. polylepis*, are almost twice as large as their mates. These species feed largely on reptiles, including limbless lizards, typhlopid snakes and amphisbaenians. Recent studies of DNA suggest that, as currently conceived, *Amblyodipsas* is not a natural lineage, with some species perhaps being more closely related to *Macrelaps* than to other species of *Amblyodipsas*, so there might be changes to the classification of some of these snakes in the near future.

A group of five unusual-looking quill-snouted snakes (*Xenocalamus*) derive their name from the resemblance of the prominent snout to that of a quill pen. These species have very slender bodies with flattened heads, minute eyes and acutely pointed snouts with a conspicuously enlarged rostral scale. They occur mostly in sandy regions and feed almost exclusively on amphisbaenians but also some burrowing snakes. Attractively marked with alternating stripes of yellow and black, two species of *Chilorhinophis* are semi-burrowing forms that also live largely on a diet of amphisbaenians. The tails of these snakes are coloured and shaped much like the head, and are used in distracting predators. *Polemon* is the most species-rich aparallactine genus, with 14 currently recognized species of 'snake eaters' that typically feed on snakes and other elongate, often burrowing prey such as caecilian amphibians.

ABOVE Quill-snouted snake, the aparallactine atractaspidid *Xenocalamus bicolor*, from southern Africa.

FAMILY MICRELAPIDAE

Micrelapidae was named as a distinct family only in 2023, to recognize the distinct evolutionary lineage comprising the three Arabian and (North)east African species of *Micrelaps* and the poorly known Somalian *Brachyophis revoili*. Previously, the relationships of these four species were not well-understood; for example *Brachyophis* was usually classified as an aparallactine atractaspidid. DNA evidence clearly demonstrates that *Micrelaps* is distinct from all other elapoid family lineages and, although DNA data are not yet available for *Brachyophis*, the only known species of that genus shares a detail of the palate bones with *Micrelaps* that suggests it is a close relative. Micrelapids are small snakes, less than half a metre (approximately 1.5 ft) long. There is little ecological data available for the family but, where known, they are oviparous and (semi-)fossorial snakes living in shrubland and savanna habitats, where they feed on arthropods and small scolecophidian snakes.

ABOVE The micrelapid *Micrelaps vaillanti*, an example from Tanzania.

FAMILY LAMPROPHIIDAE: African House Snakes, File Snakes and Water Snakes

Of the 18 genera and approximately 90 species that are currently assigned to the family Lamprophiidae, some 30 species of African house snakes (*Lamprophis, Boaedon* and *Pseudoboodon*) are small, 60–90 cm (24–36 in) long, nocturnal, constricting snakes that feed mainly on rodents and lizards. With smooth, shiny body scales in numerous rows, fairly short tails, and eyes with vertically elliptic pupils, they look superficially rather like hatchling pythons, although they differ in scale pattern. They often visit buildings in search of prey and are useful in controlling vermin. Fifteen species of African file snakes (in the genera *Goniotophis, Gracililima, Limaformosa,* and *Mehelya*), which are only distantly related to Asian and Australasian file snakes (*Acrochordus,* see p.96), eat mainly snakes, including venomous species such as cobras and night adders, although they will also consume lizards, toads and other ectothermic animals. Their common name alludes to the shape of the body, which is triangular in cross-section and resembles a three-

ABOVE A Tanzanian specimen of the lamprophiid, *Lycophidion ornatum.*

cornered file, the abdomen being rather flat and the spine raised into a prominent ridge. The body scales of these snakes are also heavily keeled and, unlike those of most colubrids, are contiguous rather than overlapping. In common with other lamprophiids, they are egg-layers. Twenty-three species of *Lycophidion*, African wolf snakes, are small snakes (typically less than 0.5 m, 1.6 ft long, and often reaching only a little more than half that length) found in sub-Saharan Africa. These are inconspicuous, mostly nocturnal, ground-dwelling snakes, many of which seem to specialize in feeding on lizards. Most of the nine species of *Lycodonomorphus* are known informally as African water snakes. With the exception of the terrestrial South African olive ground snake, *L. inornatus*, members of this genus are largely aquatic and feed on frogs and fish.

FAMILY PROSYMNIDAE: Shovel-Snout Snakes

Eighteen species, all in the genus *Prosymna*, are distributed across sub-Saharan Africa. These are small snakes, with many species not exceeding 30 cm (12 in). They burrow in loose, generally sandy soils in savanna, semi-desert and woodland habitats. Their burrowing is aided by flattened snouts that end in a sharp, horizontal anterior edge. They specialize in feeding on lizard and snake eggs, which they puncture using specially adapted, blade-like teeth in the upper jaw. As far as is known, they are all oviparous. Although considered (semi-)fossorial, *Prosymna* have also been caught in pitfall traps, so they might venture onto the surface more than is generally appreciated. Their relatively large eyes and often bold speckled, spotted, barred or lines patterns are also consistent with some above-ground activity.

FAMILY PSAMMOPHIIDAE: Sand Snakes, Skaapstekers, Bark Snakes and Beaked Snakes

This family includes eight genera of mostly very slender snakes from Africa, Mediterranean Europe and Asia. They have large eyes and prey mostly on lizards and small rodents by sight, often hunting them down at great speed during daylight.

More than 30 species of sand snakes (*Psammophis*) and six of skaapstekers (*Psammophylax*) are essentially African in distribution, with one species (*Psammophis condanarus*) in southern Asia and another (*P. lineolatus*) ranging into western China. On a drop-for-drop basis, some species have relatively potent venoms, although the quantity expelled is normally too small to cause humans serious illness. However, bites from the East African link-marked sand snakes, *Psammophis biseriatus*, and southern African spotted skaapstekers, *Psammophylax*

ABOVE Olive or hissing sand snake, the psammophiid *Psammophis sibilans*, a large sand snake found throughout much of sub-Saharan Africa in open grassland or bush country.

rhombeatus, have occasionally resulted in unpleasant envenomation in humans. Other psammophines from Africa include four species of bark snakes (*Hemirhagerrhis*), which are small and rather secretive tree-dwellers, and four species of terrestrial beaked snakes (*Rhamphiophis*), named for their sharply angled snouts.

A large and formidable species that ranges into Mediterranean Europe is the Montpellier snake, *Malpolon monspessulanus*, a 1.8 m (6 ft) long steel-grey snake capable of inflicting venomous bites that have caused clinically significant (but not life-threatening) neurological symptoms in humans. Montpellier snakes and some sand snakes have the habit of smearing themselves with a colourless liquid secreted from glands in the snout, especially after sloughing or feeding. The precise function of this 'rubbing behaviour' is not entirely clear, but because these snakes often live in dry, sun-scorched areas and are sometimes active even during the heat of midday, it may serve to reduce water loss.

FAMILY PSEUDOXYRHOPHIIDAE: Madagascan Elapoids

The majority of snakes found on the remarkable island of Madagascar are members of the gemsnake family, Pseudoxyrhophiidae. Most of the almost 90 species of this family are found only on Madagascar, the exceptions being four snail-eating species from southern Africa (*Duberria*, see p.101), Günther's racer, *Ditypophis vivax*, which is an endemic to the island of Socotra off the Horn of Africa, and three species from the Comoro Islands off the coast of Mozambique. DNA analyses suggest that today's diversity of pseudoxyrophiids on Madagascar arose from a single overseas colonisation event from mainland Africa by an opisthoglyphous (see p.21), ground-dwelling lineage, approximately 24–18 million years ago. Since then, this lineage underwent evolutionary diversification into arboreal, semi-aquatic, burrowing and other ground-dwelling species, both opisthoglyphous and aglyphous. Madagascan pseudoxyrhophiids also live in a very wide range of habitats, including rainforest, heathland, grassland and dry scrub, and they feed on diverse prey such as other snakes, lizards, frogs, birds, mammals, and occasionally invertebrates. Other than four boids, eleven

RIGHT The pseudoxyrhophiid (gemsnake) *Pseudoxyrhopus tritaeniatus* from Madagascar.

scolecophidians and two psammophiids, all living snake species on Madagascar are pseudoxyrhophiids, and it seems likely that their great diversification that took place there over the last 20 million years or so was possible because there were few other types of snake to compete with for the various available ecological niches. Although relatively little is known about the detailed feeding habits of many pseudoxyrhophiids, members of the terrestrial and sometimes burrowing genus *Pseudoxyrhopus* seem to specialize in reptile eggs. The stomach contents of a few individuals were found to contain skink and chameleon eggs, all with an intriguing slit down their sides. This may be because reptile eggs are difficult to digest unless the shell is broken, and it has been suggested that the rear fangs of some species may be important for the exploitation of this resource. Feeding on reptile eggs is a relatively well-known practice in burrowing elapids of the genus *Simoselaps*, whose dentition is also highly specialized for this purpose.

FAMILY PSEUDASPIDIDAE: Mole Snakes, Mock Vipers and the Western Keeled Snake

The pseudaspidids are a small family consisting of only four known species with a rather strange and patchy distribution, occurring from southern Africa to Indonesia, but with no members found between Africa and Nepal. The latest DNA analyses only tentatively support the grouping of these four species in a single lineage. Mole snakes, *Pseudaspis cana*, are perhaps the best-known species in the family, occurring throughout much of southern Africa where they are the largest non-venomous snake throughout their range after the pythons, growing to 2.1 m (7 ft). They are viviparous and get their common name due to them spending much of their lives living and hunting in mammal burrows. Their thick 'neck' and narrow head, with a pointed snout, likely aids them in hunting and moving in subterranean environments. The western keeled snake, *Pythonodipsas carinata*, is a desert specialist and has a small distribution in southwestern Angola and western Namibia. Its body form is almost the opposite of that seen in the mole snake with a narrow 'neck' and a relatively large head.

The two species of mock vipers, *Psammodynastes pulverulentus* and *P. pictus*, occur in South and Southeast Asia. The common mock viper, *P. pulverulentus*, has a very large distribution occupying a variety of habitats from Nepal to the Philippines and feeding on reptiles and amphibians. The common name for members of the genus come from the species looking like small pit vipers, even having enlarged rear teeth which they use to subdue prey by passing venom into them by chewing.

FAMILY CYCLOCORIDAE: Secretive Philippine Elapoids

The cyclocorids comprise a family of snakes endemic to the Philippines, where they occur across all of the major islands and many smaller islands. Despite being widely distributed throughout the archipelago, little is known about the secretive members of this family, which currently includes eight species in five genera. This family was discovered to represent a major, distinct lineage of the snake evolutionary tree only as recently as 2017. Many of the species are known only from a handful of specimens and most species are very small, only growing to < 40 cm (16 in). At a maximum known length of a little over 17 cm (7 in), *Levitonius mirus* is perhaps the smallest member of Elapoidea. Cyclocorids burrow in leaf litter and loose soil and have been found to feed on other snakes, earthworms, lizards, and eggs. The two members of the genus *Cyclocorus* have an arched maxilla bearing a couple of enlarged, fang-like teeth, which seems to be an adaptation for feeding on skinks and perhaps other reptiles.

COLUBRIDAE: Racers and garter, rat, cat, tree, reed and water snakes and relatives

Until fairly recently, the family Colubridae contained all 'higher' or 'advanced' snakes (Caenophidia) that were left over after the most distinctive groups had been placed in their own families (Viperidae, Elapidae etc.). Snake biologists understood that this was unsatisfactory, but they were somewhat overwhelmed by the morphological and ecological diversity of these many hundreds of species, and so put up with the situation while continuing their research. More recently, DNA data have been combined with a wealth of existing information on anatomy, and a clearer, more natural classification is beginning to emerge. Additional groups have been demarcated as families separate to the Colubridae (e.g., Homalopsidae, Lamprophiidae, Psammophiidae, Cyclocoridae) and there is much more certainty that those snakes now remaining in the Colubridae represent a single evolutionary lineage, with approximately 2,100 species currently recognized. Although snake biologists are now able to recognize multiple major natural lineages (see p.53) within this larger group, there are still debates as to whether this is best considered a single family (Colubridae) with eight subfamilies, or instead a single superfamily (Colubroidea) containing eight families. To a large extent this is an arbitrary decision. In this book we treat the group as a single family with eight subfamilies.

LEFT Although most colubrids are egg-layers, the European and western Asian smooth snake, *Coronella austriaca*, is among a small number of species that give birth.

BELOW The neotropical colubrine colubrid *Leptophis ahaetulla*, from Costa Rica.

Colubrids are found on all continents except Antarctica and form the main element of snake faunas in many places. It is only in Australia, where the elapids predominate, and Africa and Madagascar, where lamprophiids, psammophiids, pseudoxyrhophiids and atractaspidids are the main radiations, that colubrids are clearly in the minority. All colubrid species lack a pelvic girdle and coronoid bone (a small bone in the lower jaw primitively retained by some snakes), and many lack a left lung. Some are aglyphous, with unmodified teeth, while others are opisthoglyphous, with enlarged, grooved maxillary fangs towards the back of the upper jaw (see p.21), often connected to a venom-producing gland (Duvernoy's gland) in the mouth. Usually, the venom of these species is lethal only to the animals they feed upon, though there are a few whose bites have occasionally resulted in human deaths (see p.181). In size, colubrids range from diminutive centipede-eating snakes (genera *Tantilla* and *Tantillita*) little longer than a pencil, to the 3.8 m (12½ ft) long Asian keeled rat snake (*Ptyas carinata*) and, although many are generalist predators that feed on a wide variety of different prey, others have special adaptations for highly restricted diets. The latest DNA-based studies recognize seven or eight major lineages (here considered as subfamilies) within Colubridae. Some of these subfamilies are very small, such as the 12 species in the Scaphiodontophiinae (two species of the neotropical genus *Scaphiodontophis*, nine species of the Asian *Sibynophis*, and a single species of the very poorly known *Colubroelaps* from Vietnam), the four species in the Grayiinae (semi-aquatic African snakes in the genus *Grayia*), and the 10 species in the Pseudoxenodontinae (in the East and Southeast Asian genera *Plagiopholis* and *Pseudoxenodon*) and are not considered here in any detail. We focus instead on the four major colubrid subfamilies (Colubrinae, Natricinae, Dipsadinae and Calamariinae). This still covers more than 2,000 species of great diversity, so we are able to discuss only a small number of them in this book. In making this selection, however, we feature a broad cross-section of species.

SUBFAMILY COLUBRINAE: Rat and Tree Snakes, Racers, King Snakes and Allies

Sometimes referred to as 'typical snakes', rat snakes, racers and king snakes are among a diverse mixture of genera within the subfamily Colubrinae. Some of the common names for these colubrine snakes, especially 'rat snakes' and 'racers', have been applied to various species from different continents that might not be especially closely related, and this can cause confusion. For example, although species of the genus *Coluber* covered here (see p.165) are known as

LEFT Tiger rat snake, *Spilotes pullatus*. Although primarily an inhabitant of forests, this large colubrine colubrid species from tropical America also thrives around farms and rural settlements, where it feeds on vermin, and domestic fowl and their eggs.

racers, the same common name is applied often to superficially similar snakes in the Caribbean and South America that are classified in separate genera in a different colubrid subfamily (Dipsadinae, see p.176). There are two main lineages of colubrines, one comprising the five genera (76 species) of Asian vine snakes, gliding snakes and bronzebacks, and the other the remaining 790 species. In some classifications these two groups are both considered to have the status of subfamily (Ahaetullinae and Colubrinae, respectively).

Rat snakes and king snakes

Among the largest colubrines are two species of Asian rat snakes, *Ptyas carinatus* and *P. mucosus*, both of which are reported to exceed lengths of 3.5 m (11½ ft), and there are others that also grow to considerable sizes. The indigo snakes *Drymarchon corais* and *D. melanurus*, in particular, may occasionally exceed 3 m (10 ft) and can be almost as robust as some boas and pythons. These generalist predators eat a wide variety of vertebrate prey, especially other snakes, and they have a voracious appetite. For example, an almost 3 m (10 ft) long indigo snake from Guatemala that was caught in the act of swallowing a 1.6 m (5¼ ft) long common boa, *Boa constrictor*, already had a full-grown jumping pit viper, *Metlapilcoatlus nummifer*, itself almost 1 m (3¼ ft) in length, in its stomach.

One group of colubrines constrict their prey and occur mostly in northern temperate regions. This includes 26 species of king and milk snakes (*Lampropeltis*) and seven species of gopher snakes (*Pituophis*) from the Americas, as well as various rat snakes (e.g., *Bogertophis*, *Elaphe*, *Senticolis*). Many of these feed on rodents, although their diets typically include a wide range of other prey. King snakes will often eat other snakes, including venomous vipers and elapids such as rattlesnakes, copperheads, and coral snakes.

ABOVE From Mediterranean Europe the leopard snake, *Zamenis situla*, is a colourful colubrine colubrid of olive groves, woods, and rocky hillsides. It feeds on lizards and small mammals.

LEFT The North American colubrine colubrid, *Pituophis catenifer*, a specimen from Cedros Island, Mexico.

Racers

Noted for their agility and speed of movement are various genera of slender-bodied colubrines collectively known as 'racers'. The Old and New World colubrine racers are not all especially closely related to each other, but we treat them together here because of their similar appearance and natural history. The widespread and often locally abundant eastern racer, *Coluber constrictor*, from North and northern Central America comprises at least six distinct populations that might be considered subspecies. This species is closely related to the coachwhip, *Masticophis flagellum*, another widespread North American species with distinct, possibly subspecific populations. Racers in Central to South America include the 16 or so species of *Drymoluber* and *Mastigodryas* and 15 species of forest racers, *Dendrophidion*. In the Old World, a diversity of mostly Eurasian (though partly northeast African) racers comprise more than 75 species, most-richly classified in the genera *Platyceps* (30 species) and *Eirenis* (23 species). The cliff racer, *P. rhodorachis*, of North Africa and the Middle East is an agile climber known to scale trees and thatched roofs in search of eggs; it will also venture into houses during the day in search of house geckos, birds and mice. Some of the Eurasian racers are more commonly called whip snakes, including the 11 species in the genera *Dolichophis*, *Hierophis* and *Hemorrhois*. Racers typically hunt by day. Many have conspicuously large eyes and good eyesight, and many move around in an

ABOVE The coachwhip, *Masticophis flagellum*, a ground-dwelling colubrine colubrid from southern USA and northern Mexico.

RIGHT The slender body and large eyes with round pupils of this Central American racer, the colubrine colubrid *Mastigodryas melanolomus*, are characteristic of a fast-moving, diurnal species that locate prey by sight, in this case lizards and other small vertebrates.

alert manner with their heads held off the ground. They feed chiefly on lizards and frogs, but larger species will also eat small mammals, birds and other snakes, and some feed on invertebrates. Where known, they are all oviparous. Although primarily ground dwelling and living in relatively open habitats, colubrine racers may be found high above the ground in trees, and some are adept at climbing.

Tree snakes

Among many colubrines that have enlarged fangs in the rear of the mouth and immobilize prey by envenomation, members of three genera in particular are capable of inflicting potentially lethal bites to humans (see p.181). Vine snakes (*Oxybelis* from tropical America, *Ahaetulla* from Asia) and twig snakes (*Thelotornis* from Africa) include approximately 40 species of day-active snakes that spend almost all their lives in the trees and descend to the ground only rarely. They typically have long, exceptionally slender bodies and narrow, pointed heads with large eyes, and most species have good eyesight. Other rear-fanged, day-active, tree-living colubrines include five species of Asian flying snakes (*Chrysopelea*), which escape from predators in the treetops by launching themselves off a branch and descending to the ground or a lower branch in a controlled glide. Asian flying snakes achieve this gliding motion by expanding and flattening their rib cage for the entire length of their body (much like how cobras open their hoods) and effectively swim through the air.

Nocturnal rear-fanged colubrines include nearly 40 species of Old World tree snakes (sometimes called 'cat snakes') of the genera *Boiga* and *Toxicodryas* that have slender bodies, flattened from side to side, and large eyes with vertically elliptic pupils. Essentially tree-dwellers, some grow to impressive sizes. In particular, *B. cynodon* from Southeast Asia may attain lengths over 2.75 m (9 ft). The brown cat snake, *B. irregularis,* is one of the most notoriously destructive invasive snakes,

LEFT Asian vine snakes, colubrine colubrids of the genus *Ahaetulla*, have keyhole-shaped pupils, binocular vision, and a long grooved snout that likely assists their view

BELOW Green tree snake, *Boiga cyanea*, a rear-fanged arboreal colubrine colubrid from Southeast Asia.

decimating native birds on the Pacific island of Guam (see p.169). A striking, variable but often black-and-yellow-banded species from Southeast Asia, the mangrove snake *B. dendrophila,* is relatively stout-bodied and also has a broader diet than most of its relatives, feeding on bats, birds and their eggs, lizards, frogs, other snakes and even fish. However, this widespread snake might actually be multiple, superficially similar species and so some of this apparent breadth of diet might require a reassessment once the taxonomy is better understood. Bites from some species of *Boiga* can lead to clinically significant symptoms in humans. Other nocturnal rear-fanged 'cat' snakes include more than a dozen species of *Telescopus,* among which are the African tiger snakes, *T. beetzii* and *T. semiannulatus,* which are strikingly marked yellow-and-black-spotted snakes that feed mostly on geckos and other lizards, but will sometimes climb into trees to rob birds' nests of their eggs and nestlings.

Other colubrines

There are about 90 species of kukri snake in the South and East Asian genus *Oligodon.* These snakes are named after the Nepalese dagger because of their curved rear fangs. They are relatively small, less than 1 m (3¼ ft), mostly nocturnal and oviparous, and many are known to eat eggs. Some *Oligodon* species have brightly coloured undersides and perhaps mimic Asian elapid coral snakes (*Calliophis* species). On Taiwan (China), *O. formosanus* has been observed preying on green sea turtle nests in such large numbers that fights frequently break out, with many individuals having bitten or even partly missing tails. These *O. formosanus* show true territoriality otherwise generally unknown for snakes, and females in particular burrow into and try to defend single turtle nests from other snakes so that they can feed on the eggs over several weeks.

Also found in Central and Southeast Asia, as well as Japan, are more than 70 species of wolf-toothed snakes in the genus *Lycodon.* The genus name comes from their distinctive dentition, with three to six fang-like anterior teeth separated by a toothless space from 7 to 15 posterior teeth, the final few of which are much longer than those before them. Many species display bands of darker and lighter pigmentation that are particularly pronounced in juveniles but fade into adulthood. Some, such as the Malayan banded wolf-toothed snake, *L. subcinctus,* resemble venomous kraits such as the Malayan krait, *Bungarus candidus.* Feeding typically on lizards and frogs, *Lycodon* are oviparous and generally nocturnal and terrestrial, though some are also competent climbers. A large proportion of known species have been discovered in the past 15 years or so.

ABOVE Brown tree snake, *Boiga irregularis*.

In the 1960s, biologists began to notice a decline in native bird populations on the small pacific island of Guam. By 1987, all 10 species of birds that inhabited the island's forests were in serious trouble. Native bats and lizards had also declined drastically. It eventually became clear that the catastrophe was attributable to the brown tree snake (the colubrine colubrid *Boiga irregularis*). Indigenous to northeastern Australia, New Guinea and adjacent islands, this 3 m (10 ft) long, venomous species had been accidentally introduced to Guam on cargo shipments shortly after the Second World War. They flourished and by the 1980s were common across the island. Raiding chicken farms and causing power failures by climbing onto overhead cables, the snakes also stressed the island's economy, and several young children bitten by brown tree snakes while sleeping had suffered clinical (though not life-threatening) symptoms.

That a snake was capable of establishing itself in such numbers as to almost destroy the bird population of an entire island was initially difficult for many to comprehend. But the story is not unique. Many snake species have been introduced to ecosystems where they are not native and, for some, the conditions allow their numbers to boom. This can be due to a lack of predators and diseases that naturally control their population, or because local species are not equipped to deal with these new predators. These 'invasive' species can damage local ecosystems, or human property and health. Island ecosystems are particularly at risk of invasive species (see p.74).

The common wolf snake (the colubrine colubrid *Lycodon capucinus*) was introduced to Christmas Island, off the northwest coast of Australia, in the 1980s and it is thought to have caused the extinction of four native lizard species. Another famously invasive snake is the Burmese python, *Python bivittatus,* in Florida, USA. This native of Southeast Asia was introduced through the dumping of unwanted pets, as well as escapees from a breeding facility that was damaged in a hurricane in the 1990s. Their numbers have rocketed ever since. They are causing particular damage in the Everglades National Park where their population is estimated to be in the hundreds of thousands, and they are wreaking havoc on local mammal species.

One of the most successful invasives worldwide is the tiny flowerpot snake (the scolecophidian *Indotyphlops braminus*). As its name suggests, this is a soil-dwelling snake, and it is thought to have been accidentally spread around the world in shipments of garden and crop plants. Originally from South Asia, it now occurs on every continent apart from Antarctica. This female-only species is parthenogenetic, producing clonal offspring without needing to mate, which likely improves its ability to disperse to and colonise new ecosystems.

Control of invasive snakes usually involves trapping and culling. On Guam, dogs have been trained to sniff out brown tree snakes, and scientists are devising snake-proof bird nesting boxes. In the Florida Everglades financial incentives are given to hunters who kill and bring in Burmese pythons, including in the large annual 'Florida Python Challenge' where a cash prize of several thousand dollars is awarded for the most animals culled. Despite these programs, invasive snakes remain very difficult to control.

Found throughout the mainland neotropics, as well as on the Caribbean islands of Trinidad and St Vincent, 23 or so species of *Chironius* are large (sometimes up to 3 m, 9 ft), active, diurnal snakes known as sipos. Among the best-known members of this genus is the common sipo, *C. carinatus*, from northern South America. Like most of the other species in the same genus, it feeds on anurans and is largely arboreal, spending the nights asleep high up in trees. When disturbed this lowland inhabitant is known to raise its anterior body off the ground, hold its mouth open and hiss. If this does not deter the intruder it will puff up its neck, which it flattens like a hood to expose pale blue skin, and whip its tail back and forth. Male *C. carinatus* also partake in ritual combat during the spring, whereby they intertwine their upper bodies and each tries to get higher than their competitor. This is a behaviour it shares with the two-headed sipo, *C. bicarinatus*, and perhaps other members of the genus.

RIGHT The wolf snake, *Lycodon effraensis*, a colubrine colubrid from Southeast Asia.

SUBFAMILY NATRICINAE: Grass Snakes, Marsh Snakes, Keelbacks, Garter Snakes and Water Snakes

Approximately 270 species are currently classified in the subfamily Natricinae. They are widespread in the Old World, with three quarters of all natricine species occurring in Asia. There are also many species in the Americas, including USA, Canada, Mexico and Central America. Viviparity has evolved three times independently within the Natricinae. All but two Old World species lay eggs, whereas all species from the Americas are viviparous. Many natricines are semi-aquatic but the group as a whole also includes aquatic and burrowing forms.

From much of Europe, northwest Africa and Asia, five species in the genus *Natrix* are day-active snakes that feed mainly on amphibians and their larvae, and small fish. They are usually found in or near water, although some are more aquatic than others. In particular, the dice snake, *N. tessellata*, spends much of its time in water and often remains beneath the surface for considerable periods, but other species may be encountered in dry heathlands, meadows and woods. European *Natrix* are mostly green snakes marked with variable patterns of spots or indistinct stripes, although the viperine snake, *N. maura*, often has a zigzag pattern that increases its resemblance to a viper. By hissing fiercely, flattening its body and striking repeatedly when cornered (though usually with its mouth closed), the behaviour of this species is also convincingly viper-like. *Natrix* will also play dead to avoid predation, achieving this by rolling onto their backs, holding their mouths agape (often with their tongue hanging out), and releasing a highly pungent musk mimicking a dead, rotting snake. Other, distantly related snakes also adopt a play-dead defence (see p.51).

LEFT Grass snake, *Natrix helvetica*, a natricine occurring in western mainland Europe and the British Isles.

EGGS AND OTHER UNUSUAL DIETS

ABOVE Egg-eating snake, the colubrine colubrid *Dasypeltis scabra*, from Africa and Arabia.

Various colubrids have highly specialized feeding habits. Conspicuous among these are six species of African egg-eating colubrid snakes, *Dasypeltis*, which feed exclusively on birds' eggs and have special structures for dealing with their smooth, hard shells. On the underside of the neck vertebrae are a series of 25–35 bony spines (the hypapophyses) that project downwards like simple teeth. As the snake swallows an egg, it performs a series of sideways and downward rocking movements with its head, during which the egg is forced against these vertebral 'teeth' until it breaks, often with an audible cracking sound. Muscular contractions of the oesophagus then compress the shell and release its contents into the stomach. The snake expels the crushed empty shell shortly afterwards through its mouth. Even comparatively large eggs can be swallowed whole. A 1 m (3¼ ft) long snake with a head scarcely wider than a large fingernail, for example, is capable of consuming an average sized chicken's egg. Such incredible feats of swallowing are made possible by modifications of the head skeleton. The lower jaw is long and the ligament that connects each side at the front is highly elastic, enabling the two halves to be stretched widely apart. Inside the mouth itself, there are also loose folds of skin that lie along the lower jaw and unravel during swallowing. Among various other snakes that eat shelled eggs, the Japanese rat snake, *Elaphe climacophora*, also has spines on the underside of a few anterior vertebrae, but only the African egg-eating snakes and perhaps a poorly known Indian species, *Elachistodon westermanni* (also a colubrine colubrid), crush and regurgitate the shells.

Other interesting dietary specialisms among snakes include members of the family Pareidae that feed primarily on snails (see p.100), homalopsid species such as *Gerarda prevostiana* that eat freshly molted mangrove crabs (p.121), and the elapid sea snake *Emydocephalus annulatus* that has largely discarded its ancestor's potent venom because it specializes on eating fish eggs (p.149).

African natricines (marsh snakes)

Only 13 natricine species are known from tropical continental Africa, including six species of marsh snakes, *Natriciteres*, and various other water-dwelling species (three species of *Limnophis*, one *Afronatrix*, one *Helophis*, and two *Hydraethiops*). They feed chiefly on frogs. Some also eat fish and aquatic invertebrates. Marsh snakes are unusual among snakes in being able to break off their tails to escape from predators (known as caudal autotomy: see p.47). Among the most widespread, the African olive marsh snake, *N. olivacea*, ranges from Ghana and Sudan to Angola, Zimbabwe and southern Mozambique, in streams and marshes from sea level up to about 1,980 m (6,500 ft). Tropical continental Africa is unusual in that all of its natricines are (semi-)aquatic. The single natricine from the Indian Ocean Seychelles islands, the Seychelles wolf snake, *Lycognathophis seychellensis*, originated from an African lineage and, although it spends time close to water, is much more terrestrial.

Asian natricines (keelbacks)

Almost 200 species of natricines are known from Asia. Modern natricines originated in Asia approximately 35 to 50 million years ago, before diversifying and dispersing globally. Although Asian natricines are ecologically and taxonomically diverse, many species are collectively called 'keelbacks', especially the semiaquatic species, that live in a wide range of habitats and feed mainly on amphibians and fish. Among the most widespread, approximately nine species of *Fowlea* and five species of *Xenochrophis* are largely aquatic and commonly referred to as painted keelbacks. The chequered keelback, *F. piscator*, is a particularly common species found throughout much of southern and mainland southeastern Asia where it occurs in weed-choked ponds, slow-flowing rivers, streams, ditches and flooded rice paddies. Stream snakes (genus *Opisthotropis*) form a group of 25 species found mostly in highland areas. They generally eat soft-bodied invertebrates, amphibians and fish, though some species, such as the bicoloured stream snake *O. lateralis*, resemble north American natricines of the genus *Regina*, in feeding on crustaceans. Some Asian natricines are typically found away from water, such as the nine species of Sri Lankan *Aspidura*, which are small snakes that burrow in moist soils.

Notable among Asian keelbacks for their potentially dangerous bites are 30 species in the genus *Rhabdophis*. Most are generally mild-mannered and docile, but they have much-enlarged rear fangs and the venom of at least two species is unusually potent (see p.181). Many Asian natricines flatten their necks when alarmed, and some species of *Rhabdophis* discharge a distasteful whitish secretion at the same time from glands in the neck. The east Asian yamakagashi, *R. tigrinus*, can make this gland secretion poisonous by sequestering poisons from toads that it eats, with the amount of poison held in the glands depending on the amount of toads in the diet, making the species both venomous and poisonous. The same is possibly true of other gland-bearing species of *Rhabdophis*, but this has not yet been studied in detail.

American natricines (garter snakes and 'water' snakes)

This final group of natricines is found in the New World. Commonly referred to as water and garter snakes, they represent a single evolutionary lineage that probably originally dispersed from the Old World approximately 23 million years ago before diversifying in the Americas. These natricines are not to be confused with African 'garter snakes' (species of the elapid genus *Elapsoidea*: see p.130). Among the most widespread and common of these are some 37 species of garter and ribbon snakes, *Thamnophis*, that are largely found in the USA, although several range into southern Canada and some also live in Mexico and northern Central America. Often brightly coloured, with contrasting patterns of dorsal stripes and spots, they are small- to medium-sized, rather slender snakes with strongly keeled scales. Northwestern garter snakes (*T. ordinoides*) and western terrestrial garter snakes (*T. elegans*) in particular, are often found far from water, while others, such as western aquatic garter snakes (*T. couchii*) and narrow-headed garter snakes (*T. rufipunctatus*) are highly aquatic. Garter and ribbon snakes feed chiefly on amphibians, small fish, earthworms and aquatic invertebrates such as leeches. The common garter snake (*T. sirtalis*) sometimes preys upon the rough-skinned newt, which carries a potent neurotoxin in its skin, and it has been shown that the snakes retain the newt toxin in their livers for up to several weeks, in quantities that probably makes them poisonous to bird or mammal predators.

RIGHT Green water snake, the natricine colubrid, *Nerodia cyclopion*, largest of the North American water snakes and one of the most fecund snakes in this region. Adult females may produce more than 100 young in a single litter.

ABOVE Garter snakes are found in a wide range of different habitats but usually near pools and other bodies of freshwater. among the more thoroughly aquatic is the twin- striped garter snake, the natricine colubrid *Thamnophis hammondii*, from southern California and Mexico.

Widely distributed over large parts of eastern and southern North America, there are 10 species of highly aquatic snakes in the genus *Nerodia*. Although they are commonly referred to as water snakes, this is a name they share with many other genera across several families and geographic regions, which can be confusing. These are stout-bodied forms with strongly keeled dorsal scales and sombre-coloured markings. Females often grow considerably larger than males and those of some species, such as the green water snakes (*N. cyclopion* and *N. floridana*) and the brown water snake, *N. taxispilota*, may occasionally grow to over 1.5 m (5 ft). Water snakes are almost always found near ponds, streams, bayous, canals and lakes, particularly where there is dense aquatic vegetation and little current, although Harter's water snake (*N. harteri*) of Texas is restricted mostly to clear, swift-flowing streams and rivers, and salt-marsh snakes (*N. clarkii*) occur mostly in brackish estuaries.

Other water-living natricines from North America include the black swamp snake, *Liodytes pygaea,* and two species of crayfish snakes (*Regina*). These thoroughly aquatic species are typically found coiled among the matted roots of water hyacinth and other floating vegetation, where they feed on frogs, small fish, shrimp, crayfish and other aquatic invertebrates. Crayfish feature heavily in the diets of several species, and the queen snake (*R. septemvittata*) appears to feed almost exclusively on newly moulted crayfish whose shells have not yet hardened.

The five species of brown-and-red-bellied snakes (*Storeria*) the smooth earth snake (*Virginia valeriae*) and Kirtland's snake (*Clonophis kirtlandi*) are among a number of other North American and Mexican natricines that, although often found near water, are entirely terrestrial. These species mostly eat invertebrates and live among leaf litter in wooded areas or grass in wet meadows.

SUBFAMILY DIPSADINAE: New World Snail-Eating Snakes, Cat-eyed Snakes, Mussuranas, False Pit Vipers and Relatives

Only occurring in the New World and parts of Asia, the Dipsadinae is a taxonomically, ecologically and morphologically very diverse group of snakes that have undergone several changes in classification. Some classifications have assigned this group full family status (known as Dipsadidae or Xenodontidae) with two or three subfamilies, but here we treat it as a single subfamily within the Colubridae. With more than 830 species, the Dipsadinae is the most species-rich lineage of snakes in the New World. Dipsadines are mostly tropical, occurring in South and Central America, the Caribbean, and the Galápagos

ABOVE *Oxyrhopus rhombifer*, a rear-fanged but only mildly venomous dipsadine colubrid from tropical South America. Compare with *Micrurus surinamensis* (p.132), a more venomous elapid found in the same region.

islands (900 km; 559 miles off the Pacific coast of Ecuador), but a few species extend also into North America. In addition, three species of a single genus, *Thermophis*, occur in high elevations in eastern Asia, often in habitats warmed by natural hot springs. There are two main branches in the evolutionary tree of Dispadinae – those species more closely related to *Dipsas*, and those more closely related to *Xenodon*.

New World snail-eating snakes, cat-eyed snakes, blunt-headed snakes and night snakes: *Dipsas* and relatives

Several dipsadine genera (*Dipsas*, *Sibon*, *Sibynomorphus* and *Tropidodipsas*) share a specialization with members of the Southeast Asian family Pareidae in that they feed almost entirely on snails and slugs (see p.100). Although not closely related, both dipsadine and pareatid snail-eaters use a similar technique to remove the soft bodies of their gastropod prey, and all of these snakes are similarly small with large heads and protruding eyes. The ringed snail eater, *Geophis sartorii*, from Central America has a cylindrical body and is a ground-dwelling inhabitant of forest leaf litter, whereas the cloudy snail eater, *Sibon nebulatus*, short-faced snail

ABOVE The dipsadine colubrid, *Leptodeira septentrionalis,* sometimes feeds on egg masses of tree frogs, such as these of the red-eyed tree frog, *Agalychnis callidryas,* in Costa Rica.

ABOVE Yellow blunt-headed snake, the Neotropical dipsadine colubrid, *Imantodes inornatus*. The slender, laterally compressed body, long tail, and large eyes with vertically elliptic pupils are characteristic of snakes that live in trees and hunt by night.

eater, *Dipsas brevifacies*, and most others are adapted for climbing, with elongated bodies, flattened from side to side, with the head strongly differentiated from the slender neck.

Six species of blunt-headed snakes (*Imantodes*) are nocturnal and feed chiefly on small *Anolis* lizards, often plucking them from leaves and branches as they sleep. These snakes are extremely slender with large heads compared to the width of their bodies. Closely related, but more robustly built, are 20 species of cat-eyed snakes (*Leptodeira*). These feed on a wide range of prey, including other snakes, although they are generally considered to be mostly frog-eaters. The small-spotted cat-eyed snake, *L. septentrionalis*, eats the egg masses of leaf-breeding tree frogs, possibly being attracted by the vibrations generated by the loud calls of breeding frogs.

Several snakes on the *Dipsas* branch of the dipsadine evolutionary tree, including *Adelphicos*, *Geophis* and *Atractus*, live mostly in leaf litter and soil, chiefly on a diet of earthworms and some other soft-bodied invertebrates. With 150 species currently known and new ones frequently being discovered, the genus *Atractus* has more species than any other snake genus, and more than

ABOVE The dipsadine colubrid *Atractus tamessari* from Guyana is one of 150 species in this genus.

almost any other amniote (the group including reptiles, birds and mammals). *Atractus* range from southern Central America to the south of Brazil, and from the Pacific slopes of the Andes to the Atlantic rainforest of South America. Many species appear to have very restricted geographic ranges, particularly those found in the Andes. As is often the case with small burrowing snakes, many species of *Atractus* are known from only one or very few specimens and much remains to be discovered about their biology. Commonly known as ground snakes, they are generally small (around 20 cm, 8 in), although the giant ground snake, *A. gigas*, is much larger than the rest and can reach over 1 m (3¼ ft) long. Nine species of night snakes in the genus *Hypsiglena* live in generally arid habitats in western USA and Mexico. These are nocturnal, ground-dwelling snakes with vertical pupils and retinas seemingly specialized for night vision.

POTENTIALLY DANGEROUS COLUBRIDS

ABOVE Boomslang, the colubrine colubrid *Dispholidus typus*, in characteristic defence posture. Boomslangs are widely distributed throughout much of Africa, and are entirely arboreal in habits and mostly diurnal. They feed mainly on small birds and lizards, especially chameleons.

Many colubrids have enlarged (but not hollow) fangs in the rear of the mouth and, although much less-well studied than venom in elapids and vipers, perhaps one-quarter or more of all colubrids produce venom in mouth glands (called Duvernoy's glands). Until just over 65 years ago it was believed that all colubrid bites were generally harmless to humans. The death in 1957, however, of a prominent herpetologist following the bite of a boomslang (the African colubrine *Dispholidus typus*), set alarm bells ringing that changed this view rapidly.

People who have experienced the effects of boomslang envenomation as well as the venoms of other potentially dangerous African species such as cobras have remarked that, by comparison, the bite of the boomslang is the most painful and distressing. Among the worst of its unpleasant symptoms is profuse internal bleeding. The boomslang is generally an inoffensive snake that tends not to bite unless it is seriously provoked, but its venom is evidently potent.

Bites from various other rear-fanged colubrids have also been known to cause clinically significant symptoms in humans. In particular, those of African vine or twig snakes (species of the colubrine colubrid genus *Thelotornis*) have caused several fatalities, including the death in 1972 of an eminent German herpetologist, and more recently deaths have been recorded following bites inflicted by an Asian species of keelback (subfamily Natricinae), the yamakagashi, *Rhabdophis tigrinus*. Other colubrids that appear to have particularly potent venoms include Asian cat or tree snakes (species of the colubrine genus *Boiga*) and among the subfamily Dipsadinae, the Central American road guarder, *Conophis lineatus*, and some neotropical racers, particularly species of the genus *Philodryas*.

Despite these cases, most colubrids are not considered venomous. Even the venomous species by and large pose no special health risk to humans in the wild because envenomation is exceptionally rare and occurs almost invariably during direct handling.

Mussuranas, false pit vipers and neotropical water snakes: *Xenodon* and relatives

The *Xenodon* branch of the dipsadine evolutionary tree comprises a diverse group of mostly South American snakes. They include 10 species of mussuranas (in the closely related genera *Clelia* and *Mussurana*), renowned for their capacity to overpower and eat venomous pit vipers. Some mussuranas undergo a marked colour change with age. Juveniles are bright orange-red with a black head and pale neck ring, but when they reach about 60 cm (24 in) in length they gradually darken, and with successive moults of the skin eventually change to a uniform deep bluish-black (see p.46). Twelve species of false pit vipers (*Xenodon*) have colour patterns strikingly similar to those of some pit vipers; when provoked, these frog and toad-eating snakes recoil and hiss loudly, similar to pit vipers, and also flatten their necks. Twenty species of neotropical water snakes (*Helicops*) display several classic adaptations to aquatic life, including eyes and nostrils on the top of the head and viviparity. One member, *H. leopardinus*, is an abundant species in

ABOVE Mussurana, *Clelia clelia*, a powerful 2.5 m (8 ft) long Neotropical species that frequently predates on venomous snakes, and which is largely immune to the highly toxic venom of pit vipers.

the giant Pantanal wetland region of Brazil. Like its congeners, this species feeds largely on fish, with frogs making up a lesser but considerable part of the diet, and it actively forages in both shallow pools and at the bottom of deeper water. It is often found in association with floating vegetation, a microhabitat that supports large numbers of small fish. The false coral snake genus *Erthryolamprus* includes 55 ecologically diverse species, several of which are also commonly found in wet lowland areas.

Other South American dipsadines on the *Xenodon* branch of the evolutionary tree include two species of false water cobras, *Hydrodynastes*, and 16 fast-moving snakes in the genus *Philodryas*. One semi-arboreal species from Brazil's Atlantic Forest, *Tropidodryas striaticeps,* is unusual among colubrids in having a long, thin prehensile tail. Juveniles of this species have a yellow-white tail with flared scales and they have been observed undulating the tip as a lure to attract lizards and other prey items. This species also actively forages, and larger prey items such as rodents are often constricted, particularly by adults, although they are probably also subdued with the use of venom.

One lineage of dipsadine snakes dispersed from the South American mainland and into the Caribbean. Based on DNA evidence it seems to have diversified rapidly over the last 10 million years into some 50 species in 11 genera. These West Indian 'racers' are now classified in their own tribe, Alsophiini, which constitute almost the entire diversity of colubrids in this part of the world. Most West Indian racers are ground-dwelling and brownish, like their probable South American ancestor, but three species of the genus *Uromacer* on the island of Hispaniola are green and arboreal. At lengths of up to almost 3 m (10 ft) the Hispaniolan brown racer, *Haitiophis anomalus*, is one of the largest dipsadines; its defensive display includes a flattened cobra-like neck. Island species often face strong conservation threats, and the Jamaican *Hypsirhynchus ater* has not been seen in more than 90 years, so is possibly extinct.

The West Indian racers were thought to be closely related to superficially similar snakes of the Galápagos islands, but information on hemipenis morphology and DNA has confirmed that the Galápagos species are a separate radiation and not especially closely related. All nine currently known species of Galápagos snake are members of the genus *Pseudalsophis*, with the only other species in this genus (*P. elegans*) occurring in Peru, Ecuador and Chile on the South American mainland. Based on rates of DNA evolution (calibrated by the fossil record), these dipsadine 'racers' reached the Galápagos in a single dispersal event from mainland South America approximately 4–7 million years

ABOVE False pit viper, the dipsadine colubrid *Xenodon merremi*, from South America. The enlarged fangs of this opisthoglyphous species can be seen in the rear of its mouth. Note also the extensive mouth cavity and forward-placed opening of the windpipe, modifications that enable snakes to swallow large prey.

ago, where they diversified into multiple species and further dispersed to many of the Galápagos islands in an approximately east to west direction. Today, no more than two species are found together on any single island, and then always in a combination of one larger (to more than 60 cm snout-vent length; 24 in) and one smaller (less than 40 cm snout-vent length; 16 in) species. This size difference appears to allow two species of ground-dwelling racers to co-exist in close proximity, possibly by exploiting slightly different diets and/or microhabitats. The species can often be found in great densities on the islands when prey abundance is high.

SUBFAMILY CALAMARIINAE: Asian Reed Snakes
Seven genera and about 96 species of reed snakes are grouped in the subfamily Calamariinae. These are small, only growing up to about 45 cm (1½ ft) long, shiny-scaled snakes found mainly in Southeast Asia. They are adapted for burrowing,

with slender, cylindrical bodies, few head scales and a rigidly constructed skull. If disturbed on the surface, they often wriggle down into the soil at speed. In both skin patterns and behaviour, some are superficially, but strikingly, similar to highly venomous elapids. For example, pink-headed reed snakes, (*Calamaria schlegeli*), are deep bluish-black with a bright orange-red head, like the Malayan long-glanded coral snake (*Calliophis bivirgatus*) and the red-headed krait (*Bungarus flaviceps*). All calamariines are oviparous and they eat mostly earthworms and insect larvae, although the diet of some larger species, such as *Calamaria lumbricoidea*, also includes skinks. The diminutive reed snake, *Pseudorabdion longiceps*, is a relatively common and widespread species from mainland and insular Southeast Asia. Usually less than 20 cm (8 in) long, it inhabits low-lying forests and cultivated areas such as rice paddies. Like most calamariines it is semi-fossorial and is most often encountered when looking under stones and in damp decaying plant matter. In 2023 herpetologists in Malaysia reported observing this reed snake at night escaping from potential predators by throwing itself downhill in a dramatic series of loose cartwheels. It is not yet known if other reed snakes adopt this defensive behaviour This subfamily is intriguing because so little is known about the natural history of any of its members. With eight new species of Calamariinae described in the last 10 or so years, there is much still to learn about this group.

LEFT The calamariine colubrid, *Calamaria grabowskyi*, from Borneo.

Glossary

ABDOMEN the part of the body containing the digestive organs, the belly

ADAPTIVE RADIATION the evolution of a natural group of species into a diverse range of forms adapted for different niches

AESTIVATION extended dormancy during periods of heat and/or drought

AGLYPHOUS simple dentition; lacking fangs

AMINO-ACID PEPTIDES short chains of amino acids, which are organic compounds that in longer chains form proteins

AMNIOTE a member of the Amniota – the group of vertebrate animals with multiple membranes surrounding the embryo within the egg, including living reptiles, birds and mammals

ANTIVENOM biological product for treating venomous bites, typically made by collecting antibodies produced when dilute venom is injected into domestic mammals

ANURANS members of the amphibian order Anura – frogs and toads

APOSEMATIC colour and patterning that serves to warn off predators, often advertising the dangerous nature of an animal

AQUATIC living in water

ARBOREAL living in trees

ARCHIPELAGO a group of islands

ASPHYXIATE to restrict oxygen, for example by suffocation

AUTOHAEMORRHAGE spontaneous bleeding behaviour that has evolved in some snakes (e.g., genus *Tropidophis*) probably as a means of deterring predators

AUTOTOMY shedding part of the body, usually the tail, either spontaneously or when grasped by a predator

BRILLE transparent covering to eye in most snakes, also called the spectacle

CAECILIAN a limbless, elongate tropical amphibian, a member of the group Gymnophiona

CARNIVOROUS meat-eating

CARRION dead meat eaten by scavengers

CAUDAL of the tail

CHEMORECEPTION odour detection; the senses of taste and smell

CHYME partly digested food passed from the stomach to the intestine

CLOACA the common chamber into which the reproductive and digestive tracts discharge their contents, emptying to the outside through the vent

CONGENER a species classified within the same genus

CONSTRICTION method of disabling or killing prey by coiling tightly to restrict breathing and blood flow

CONVERGENT EVOLUTION the independent acquisition of similar features by species that are not especially closely related, often in response to a similar way of life

CORONOID a small bone of the lower jaw found in some snakes that retain this primitive feature

CREPUSCULAR active shortly before dawn and/or dusk, during twilight

CYTOTOXIC action of a toxin that primarily kills cells

DENTARY the front, main tooth-bearing bone in a snake's lower jaw

DERMATITIS skin irritation

DIURNAL active during the day

DORSALS scales on the body of a snake except those along the midline of the belly (ventrals)

DORSAL or **DORSUM** the back of an animal

DUVERNOY'S GLAND venom-producing gland in the mouth of rear-fanged colubrid snakes, named after D.M. Duvernoy, a French anatomist. Duvernoy's glands are somewhat dissimilar to the venom glands of vipers, elapids and atractaspidids, but they might nonetheless have evolved from the same ancestral structure

ECTOPTERYGOID a bone in the upper jaw forming part of the roof of the mouth

ECTOTHERMIC dependent on external (environmental) conditions to regulate body temperature

ENVENOMING an act or instance of the introduction of venom into the body, such as injection via the fangs of venomous snakes

FOSSILIZATION geological preservation of the remains of organisms, producing a fossil

FOSSORIAL burrowing

GASTROINTESTINAL of the stomach and intestine

GASTROPOD a snail or slug, a member of the mollusc class Gastropoda

GLOTTIS entry to the tube (the trachea) that leads to the lungs

HAEMAPOPHYSIS small bone projecting downwards from the underside of a tail vertebra

HAEMORRHAGE excessive bleeding from damaged blood vessels

HAEMOTOXIC action of a toxin that primarily attacks the blood and circulatory system

HEMIPENIS (PL. HEMIPENES) one of the paired male copulatory organs of snakes

HERPETOLOGY the study of amphibians and reptiles, carried out by herpetologists

HIBERNATION extended dormancy during cold periods

HYBRIDISATION the crossing of two different species to create a hybrid

HYPAPOPHYSIS ventral projection on a vertebra in the body of a snake

HYPOTENSIVE SHOCK sudden low blood pressure

HYPOVENTILATE inadequate ventilation of the lungs caused by slow or shallow breathing

INFRARED electromagnetic radiation that has a wavelength slightly longer than that of visible light (but shorter than microwaves and radio waves)

INTERNASALS a typically paired set of scales on the top of a snake's snout, approximately between its nostrils

IRIDESCENCE rainbow-like light reflected from a surface that appears to move when viewed from different angles

JACOBSON'S ORGAN see vomeronasal organ

KEELED having a keel or ridge

KERATIN a group of proteins that are the main toughening components of scales, also found in claws, hair and skin

LATERITIC type of rock or soil rich in iron and aluminium that forms from prolonged weathering of rock in typically warm and moist conditions

LABIALS head scales along edges of a snake's upper and lower lips

LOREAL a scale on the side of a snake's snout between the nostril and eye, but touching neither

MACROPHAGOUS feeding on relatively large items of food

MACROSTOMOUS having a large-gaped mouth

MANDIBULAR of the lower jaw (the mandible)

MARINE living in the sea

MAXILLA a bone in a snake's upper jaw

MECHANORECEPTOR sensory receptor that detects stimuli of pressure, touch, distortion or vibration

MELANOPHORES pigment-containing cells in the skin

MENTAL GROOVE midline cleft between the scales on the underside of the chin in some snakes

METABOLIC RATE amount of energy expended by an animal when at rest

MICRO-COMPUTED TOMOGRAPHY (MICRO-CT) technique combining multiple x-rays to create three dimensional images of the inside of an object

MICROORNAMENTATION microscopic surface features, such as on scales

MICROVILLI microscopic projections on the surface of some cells that facilitate absorption and secretion

MIMICRY resemblance of one species to another, often distasteful or harmful species

MONOTYPIC lineage (e.g. genus or family) represented by only a single known species

MORPHOLOGY form, structure and appearance – or the study of those aspects of organisms

MYOTOXIC activity of toxins that destroy muscle tissue

NEOTROPICAL from the tropical regions of the New World (the Americas)

NEURAL SPINE the dorsal projection on a vertebra

NEUROTOXIC activity of toxins that have a particularly marked effect primarily on nerve tissues

NEW WORLD the Americas (sometimes used by non-biologists to also include Australia and New Zealand)

NICHE ecological role or position

NOCTURNAL active during the night

ODORANT a substance or molecule that has a smell

OESOPHAGOUS the muscular tube between the mouth and stomach

OLD WORLD the continents of the eastern hemisphere known before the 'discovery' of the Americas, i.e. Europe, Asia, and Africa

OLFACTORY relating to the sense of smell

OPHIDIOMYCOSIS a fungal disease affecting snakes

OPISTHOGLYPHOUS having fangs toward the back of the mouth; rear-fanged

OVIPAROUS egg-laying

PALEONTOLOGIST a person who studies fossils and the biology of extinct organisms

PALATINE a bone of the palate (roof of the mouth) that bears teeth in many snakes

PARASITIC lifestyle whereby one organism (parasite) benefits at the expense of another (host)

PARTHENOGENETIC development of viable offspring without fertilization from a male

PELVIC GIRDLE the hip (pelvis), connecting the hindlimb to the body

PHEROMONE biochemical produced by one organism that triggers a response in another of the same species; for example, sex pheromones convey information about species identity and reproductive condition

PHOTORECEPTION the detection of light

PHOSPHOLIPASES enzymes that break down particular fats into smaller molecules

PHYSIOLOGY the function of living organisms, also the study of this

PIT ORGAN organ used by snakes to detect infrared electromagnetic radiation (infrared light)

POLYGONAL having multiple straight sides

PREHENSILE able to grasp

PREMAXILLA a small bone on the front end of the snout that in a few snakes bears small teeth

PRIMARY BIFURCATION the first (deepest) split of one line or branch into two in a branching system, such as a tree of evolutionary relationships

PROCRYPSIS colour and patterning designed to conceal an animal in its natural habitat

PROTEROGLYPHOUS having fangs in the front of the mouth that are largely immovable

PROTUBERANCE something that sticks out from a surface, an outgrowth

PTERYGOID an often tooth-bearing bone in the back of the roof of the mouth

QUADRATE a bone in the back of the skull that articulates with the lower jaw

RADIOTELEMETRY using radio signals to determine a location

RECTILINEAR type of locomotion in which a snake moves slowly forwards in a straight line, without bending the body

REGURGITATE to bring food that has been swallowed back up into the mouth

RELICT something that has survived, typically in reduced numbers and/or smaller distribution

RETINA light-sensitive tissue at the back of the eye

ROSTRAL SCALE the scale at the tip of a snake's snout

SEXUAL DIMORPHISM differences in morphology between males and females

SIDEWINDING type of locomotion in which only small areas of a snake's body are in contact with the ground at any one time and in which the body does not move while in contact with the ground, particularly used by vipers to move on loose sand

SLOUGHING the process of shedding the outer layer of the skin

SOLENOGLYPHOUS having fangs in the front of the mouth that are hinged and erectile

SQUAMATES reptiles of the order Squamata, comprising lizards and snakes

STAPES a small bony rod at the back of the skull through which sound vibrations are transmitted to the inner ear

STERNUM the breastbone – absent in snakes

SUBCAUDALS the scales on the underside of the tail

SUPRAORBITAL BONE a bone located immediately above the eye socket

SUPRATEMPORAL a bone that links the quadrate and lower jaw assembly with the back of the skull

TAXONOMY the naming and classification of things, such as different organisms

TERRESTRIAL living on land

THERMOREGULATION the regulation of body temperature

THORAX chest and upper body

TOXIN poisonous substance

TRACHEAL LUNG an additional respiratory organ formed by part of the windpipe

TRANSVERSE crossing from side to side, at right angles to the long axis

TUBERCULES small, raised bumps, such as on the skin

TYPE LOCALITY locality from which the specimen used to first describe a new species was collected

URIC ACID (UREA) organic compounds that are the waste products of many organisms

UV (ULTRAVIOLET LIGHT) electromagnetic radiation with a wavelength shorter than visible light

VENOM substance containing toxins that is injected into prey (or attackers) by biting

VENTRALS scales along the midline of the underside of a snake's body

VESTIGIAL forming a remnant body part that has evolved from a more functional (and typically larger) ancestral structure

VIVIPAROUS giving birth to young rather than laying eggs, sometimes termed 'live-bearing'

VOMERONASAL ORGAN organ in the roof of the mouth for sensing odour, especially those odorant chemicals picked up by the flicking of the tongue

Further information

FURTHER READING

Australian snakes: a natural history, Richard Shine. Reed Books, 1991.

Boas and pythons of the world, Mark O'Shea. Princeton University Press, 2007.

The dangerous snakes of Africa, Stephen Spawls and Bill Branch. Blandford Press, 1995.

The new encyclopedia of snakes, Chris Mattison. Princeton University Press, 2007.

Homalopsid snakes: evolution in the mud, John C. Murphy. Krieger Publishing Company, 2007.

How snakes work: Structure, function and behavior of the World's snakes, Harvey B. Lillywhite, Oxford University Press, 2014.

Mean and lowly things, Kate Jackson. Harvard University Press, 2008.

Sea snakes, 2nd edn., Harold F. Heatwole. Krieger and University of New South Wales, 1999.

Secrets of the snake charmer, John C. Murphy. iUniverse, Bloomington, 2010.

Snake: The essential visual guide, Chris Mattison, Dorling Kindersley, 2016.

Snakes: A natural history, Roland Bauchot. Sterling, 2006.

Snakes of the World: A guide to every family, Mark O'Shea. Princeton University Press, 2023.

Snakes: Smithsonian answer book, George R. Zug and Carl H. Ernst. Smithsonian Institution Press, 2004.

Snakes: The evolution of mystery in nature, Harry W. Greene. University of California Press, 1997.

So many snakes, so little time, Rick Shine. Routledge, 2022.

The book of snakes: A lifesize guide to six hundred species from around the World, Mark O'Shea, Chicago University Press, 2018.

Tales of giant snakes: A historical natural history of anacondas and pythons, John C. Murphy and Robert W. Henderson. Krieger, 1997.

Tracks and shadows: Field biology as art, Harry W. Greene. University of California Press, 2013

Venomous snakes of the world, Mark O'Shea. Princeton University Press, 2011.

INTERNET RESOURCES

The Reptile Database: www.reptile-database.org
[comprehensive listing of all living reptile species with details of distribution and classification]

International Union for Conservation of Nature (IUCN) Red List of Threatened Species: https://www.iucnredlist.org/
[information on the conservation status of species of animals, fungi and plants for which this has been assessed]

World Health Organization's snake envenoming pages: https://www.who.int/health-topics/snakebite#tab=tab_1
[information about snakebite and treatment]

Index

Picture credits

p.8, 35, 55, 79, 154, 155 ©Michele Menegon; p.7, 21, 50, 75(right), 82, 84, 85, 88, 95 (top), 107, 125, 161 (top), 164 (top) 171 ©Peter Stafford; p.10 (right) ©Samuel Lalronuga; p.13 ©Jean-Claude Rage; p.14 ©Kenney Krysko/University of Florida; p.15 (left) ©Jon Gower; (right) ©graphego/Shutterstock; p.17 (top) ©Arie Van T'Riet/Science Photo Library; (bottom) ©Scott Camazine/ Science Photo Library; p.23, 28, 30, 61, 66, 78, 87, 91, 92, 134, 164 (bottom), 172, 174, 175 ©Chris Mattison; p.24 ©Danielle Bradke; p.25, 178 ©Katie Garrett; p.26 ©Simon Maddock; p.32 (left) ©Abhishek Chinnappa; (right), p.62, 141, 146 ©Steve Swanson; p.34, 56 ©Michael and Patricia Fogden/Minden/naturepl.com; p.36 ©Robert Valentic/naturepl.com; p.38, 49, 108, 114, 115, 126, 129, 137, 167 (top), 179, 184 ©Wolfgang Wüster; p.39 Tony Phelps/ naturepl.com; p.41 ©Tim Allen; p.42 ©Jiribalek/Dreamstime.com; p.44 ©Paul Lloyd/Shutterstock; p.45 (top) ©Eric Smith; (bottom) ©Saunders Drukker; p.46 ©Díaz-Ricaurte JC, Arteaga A (2021) Common Mussurana (Clelia clelia). In: Arteaga A, Bustamante L, Vieira J, Guayasamin JM (Eds) Reptiles of Ecuador: Life in the middle of the world. Available from: www.reptilesofecuador. com. DOI: 10.47051/YKSW1188; p.48 (top), 58, 76, 77, 109, 110 (top), 111, 118, 128, 135, 136, 157 ©Stephen Von Peltz; (bottom) ©Jean Paul Ferrero/ardea; p.54 ©Ansil B.R./iNaturalist; p.59, 180 ©Philippe Kok, Royal Belgian Institute of Natural Sciences; p.65, 86, 132 ©Patrick K Campbell/Shutterstock; p.67 ©Pete Oxford/ naturepl.com; p.68, 185 Photo: Indraneil Das; p.69, 100 ©Ch'ien Lee/naturepl.com; p.71, 94 ©Ashok Captain; p.72 ©Richard

Gibson; p.73 ©Wolfgang Grossman; p.74 © captainjack0000/ iNaturalist; p.75 (left) ©Johnbell/Dreamstime.com; p.80 ©Thai National Parks; p.89, 90, 95 (bottom), 142, 169 ©Mark O'Shea; p.93 ©Benny Trapp/Wikimedia Commons; p.97 ©Muftiadi Utomo/Dreamstime.com; p.98, 99 ©David Gower; p.103 ©Thomas Eimermacher; p.104 (left) ©Kartik Sunagar; (right) ©Nicholas Casewell; p.105 (top) ©Steve Spawls; (bottom) ©Gus Benson; p.106 ©Thomas Eimermacher; p.110 (bottom)©Matthijs Kuijpers/ Dreamstime.com; p.112 ©Mirkorosenau/Dreamstime.com; p.113 ©Jonathan Kolby; p.116 ©Doug Wechsler/naturepl.com; p.119 ©John C.Murphy, JCM Natural History Photography; p.120 ©Matthijs Kuijpers/Dreamstime.com; p.123 ©Christian Ching; p.127 ©NickEvansKZN/Shutterstock; p.131 ©Stu Porter; p.139 ©Ken Griffiths/Shutterstock; p.140 ©Jiri Lochman/naturepl.com; p.144 ©David Fleetham/naturepl.com; p.148 ©Jacob Loyacano; p.150, 153 ©Minden Pictures/Superstock; p.158 ©M.Vences; p.161 (bottom) ©Daniel Jara/iStock; p.163 ©Pete Oxford/naturepl.com; p.165 ©IrinaK/Shutterstock; 167 (bottom) ©slowmotiongli/iStock; p.170 ©Zdeněk Mačát/Dreamstime.com; p.177©Michael D Kern/ naturepl.com; p.181 ©Steve Spawls.

Unless otherwise stated, images are copyright of the Trustees of the Natural History Museum, London. Every effort has been made to contact and accurately credit all copyright holders. If we have been unsuccessful. We apologise and welcome correction for future editions.

Authors' acknowledgements

We thank the following people for expert reviews, advice, information and encouragement while preparing this new edition: Trudy Brannan, Sharon Brook, Nicholas Casewell, Jenna Crowe-Riddell, Maddy Fowler, Jonathan Kolby, Michele Menegon, Kate Sanders, Kay Saunders,

Gemma Simmons, Anna Smith, Kartuk Sunagar and Jeff Streicher. We thank the skilled fieldworkers, artists and photographers who provided images for use in this book, and also acknowledge the contributions of the people additionally thanked in the previous two editions.